Lyman Alpha Emitting Galaxies at High Redshift:
Direct Detection of Young Galaxies in a Young Universe

by

Steven Arthur Dawson

ISBN: 1-58112-294-2

DISSERTATION.COM

Boca Raton, Florida
USA • 2005

Lyman Alpha Emitting Galaxies at High Redshift:
Direct Detection of Young Galaxies in a Young Universe

Dissertation.com
Boca Raton, Florida
USA • 2005

ISBN: 1-58112- 294-2

Lyman Alpha Emitting Galaxies at High Redshift: Direct Detection of
Young Galaxies in a Young Universe

by

Steven Arthur Dawson

B.A. (University of California at Irvine) 1998
B.S. (University of California at Irvine) 1998
M.A. (University of California at Berkeley) 2000

A dissertation submitted in partial satisfaction of the
requirements for the degree of
Doctor of Philosophy

in

Astrophysics

in the

GRADUATE DIVISION
of the
UNIVERSITY OF CALIFORNIA, BERKELEY

Committee in charge:
Professor Hyron Spinrad, Chair
Professor Chung–Pei Ma
Professor William L. Holzapfel

Fall 2005

The dissertation of Steven Arthur Dawson is approved:

_____

Chair   Date

_____

Date

_____

Date

University of California, Berkeley

Fall 2005

Lyman Alpha Emitting Galaxies at High Redshift: Direct Detection of Young Galaxies in a Young Universe

Copyright 2005

by

Steven Arthur Dawson

Abstract

Lyman Alpha Emitting Galaxies at High Redshift: Direct Detection of Young

Galaxies in a Young Universe

by

Steven Arthur Dawson

Doctor of Philosophy in Astrophysics

University of California, Berkeley

Professor Hyron Spinrad, Chair

An early result of galaxy formation theory was the prediction that the copious ionizing radiation produced in nascent galaxies undergoing their first starbursts should in turn produce a strong Lyα emission line. We report on our efforts to detect and characterize primeval galaxies by searching for this expected Lyα signature with two observational techniques: serendipitous slit spectroscopy, and narrowband imaging selection. In Part I, we describe our serendipitous slit spectroscopy survey of the Hubble Deep Field and its environs, which resulted in a catalog of 74 spectroscopic redshifts spanning $0.10 < z < 5.77$, including a galaxy cluster at $z = 0.85$ and five galaxies at $z > 5$. Follow–up observations at higher resolution resulted in the additional serendipitous detection of a strong Lyα–emitting galaxy at $z = 5.190$ (ES1). At the time of its discovery, ES1 was one of only nine known galaxies at $z > 5$, and was the sixth most distant known galaxy. The unprecedented spectral purity of the observation offers evidence for a galaxy–scale outflow with a velocity of $v > 300$ km s^{-1}, consistent with wind speeds observed in powerful local starbursts (typically 10^2 to 10^3 km s^{-1}), and with simulations of the late–stage evolution of Lyα emission in star–forming systems. Our final serendipitous detection is the remarkable source CXOHDFN J123635.6+621424, which is both the highest redshift known spiral galaxy, and a rare example of a high redshift, hard X–ray–emitting Type II AGN. Significantly, all of these results were acquired with no direct allocation of telescope time.

In Part II, we report on our implementation of narrowband imaging selection, with which we traded redshift coverage for survey volume, focusing on the systematic study of galaxies at a particular epoch in favor of chasing that rare, most–distant object. This effort resulted in a catalog of 76 $z \approx 4.5$ Lyα–emitting galaxies spectroscopically–confirmed in campaigns of Keck/LRIS and Keck/DEIMOS follow–up observations to candidates selected in the Large Area Lyman Alpha (LALA) survey. We find that a significant fraction of the confirmed Lyα lines have rest–frame equivalent widths ($W_\lambda^{\mathrm{rest}}$) which exceed the maximum predicted for normal stellar populations: $12\% - 27\%$ of the galaxies in the sample show $W_\lambda^{\mathrm{rest}} > 240$ Å (90% confidence), and $17\% - 31\%$ show $W_\lambda^{\mathrm{rest}} > 190$ Å (93% confidence). Furthermore, the narrow velocity widths ($\Delta v < 500$ km s^{-1}) together with a lack of high–ionization state emission lines support the conclusion that the Lyα emission in these sources derives from star formation, not from AGN activity. However, the nondetection of He II λ1640 in both individual and composite spectra (to a 2σ [3σ] upper limit of 13% [20%] of the flux in the Lyα line) dictates that though these galaxies are young, they show no evidence of being truly primitive, Population III objects. We also find that the Lyα lines in this sample systematically disfavor combinations of low velocity width (< 150 km s^{-1}) and high luminosity ($> 5 \times 10^{42}$ erg s^{-1}), suggesting that more than dust content alone, it is the velocity structure of a galaxy that determines the detectability of Lyα in emission. Finally, we construct a luminosity function of $z \approx 4.5$ Lyα emission lines for comparison to a set of Lyα luminosity functions spanning $3.1 < z < 6.6$. We conclude that if there is evolution in the Lyα luminosity function over these epochs, its significance is below the statistical uncertainty of these data. This result supports the conclusion from several smaller samples of high–redshift Lyα–emitters that the intergalactic medium remains largely reionized from the local universe out to $z \approx 6.5$. However, it is somewhat at odds with the pronounced drop in the cosmic star formation rate density recently measured between $z \sim 3$ and $z \sim 6$ in Lyman–break galaxies, and therefore potentially sheds light on the relationship between the two populations.

Professor Hyron Spinrad
Dissertation Committee Chair

to the current Dr. Dawson, the future Dr. Dawson,
and the Mrs. Dawson who made it all possible

Contents

List of Figures

List of Tables

Acknowledgments

I owe incalculable debts of gratitude to my forebears along two distinct lines of heredity: one academic, one biological. My advisor, Hy Spinrad, is the patriarch of a vast family tree of astronomers, of which I am the last of the line. I wish to thank Hy not only for the intellectual freedom he afforded me and for the scientific insights he offered, but also for allying me, as my intellectual father, to his former students, my intellectual older brothers. In ascending order: Dan Stern taught me literally everything I know about observing. Arjun Dey, with his apparently limitless erudition, taught me nearly everything else. More importantly, with major contributions from Mark Dickinson, the Spinradians taught me how to enjoy myself as a scientist, to prioritize scuba diving, record shopping, and ultimate frisbee right alongside the acquisition of data. My academic family tree also harbors a few wacky cousins, uncles, and neighbors: I benefited greatly both personally and scientifically from guidance and/or comic relief provided by Wil van Breugel, Andy Bunker, Mark Lacy, Wim de Vries, Michiel Reuland, Steve Croft, Buell Jannuzi, James Rhoads, and Sangeeta Malhotra.

I owe humble thanks of at least equal magnitude to the members of my literal family tree. My parents have been unflagging in their support of my long academic career. My admiration for the adroitness with which they match high expectations with unconditional acceptance is boundless. My sister has been a model of both industry and empathy as she pursues her own Ph.D.; thankfully, our periodic urges to chuck it all were never in phase. It's a simple fact that I would neither have pursued a Ph.D., nor stuck with it, if my father hadn't provided the model (along with occasional bursts of zen–like wisdom), my mother hadn't provided the financial, emotional, and motivational support, or if my sister hadn't provided the tiniest edge of sibling rivalry.

Finally, it is essential that I acknowledge that any shred of perspective or sanity I retained throughout this experience is owed to those folks I am so grateful to call my friends. Spanning elementary school to the present, Danny Ronen, Angie Conley, Jon Klein, Kim Decker, Liz Green, Gina Rappleye, Rob Badzey, Todd Schweisinger, Lindsey Westbrook, and Pat Garvey were separately and collectively critical in helping me to weather the tribulations of the past decade, and were indefatigable in celebrating the successes. In addition, a handful of the denizens of Campbell Hall (and their spouses) transcended mere astronomy and became great friends in their own right, namely Julie Walters, Megan Eckart, Tim Robishaw, Erin & John Johnson, and the founding members of the Astronomy Student Society (Alison Coil, Dave Sherfesee, Josh Simon, and Jon Swift). And speaking of mysterious societies, the members of the following eclectic associations provided endless distraction and affection: the Blue Bombers (Peter & Erin Marietta, Yurah & Andrew Yen, Kelly Givner, Andrea Barton–Elson, and Rachel Zimmerman); the Tower House Gang and its tangled legacy (e.g. Kris & Sammy Ahmed, and Ian Farrell), the Delicate Flower, Po', Rock Club, and the Mixturbators. And where would I be without the warm support of Sarah McCarthy, and her mixes, and her brunches? Thank you for reminding me that there is life to be had outside of this office.

And now, let's get astrophysical.

Chapter 1

Introduction: The Failed Search for Primeval Galaxies

Almost four decades ago, Partridge & Peebles (1967) predicted that galaxies undergoing their first throes of star formation should be strong emitters in the Lyα emission line. Coupled with this prediction was the expectation that the detection of such young galaxies would offer insight into theories of galaxy formation and evolution, would provide *in situ* information on star formation under conditions not found locally, and would likely lead to the detection of objects at high redshift, providing insight into physics at early epochs. Motivated by these considerations, observers spent the next 30 years searching for the highly redshifted Lyα emission which would signify the expected widespread population of galaxies undergoing their first starbursts. Though a handful of Lyα–emitters were detected in the vicinity of objects already known to be at high redshift (e.g. Djorgovski et al. 1985, 1987; Hu & McMahon 1996; Hu et al. 1996; Petitjean et al. 1996), these results were dismissed as a consequence of the possibly anomolous environments around the target objects (Cowie & Hu 1998). Meanwhile, the results of all blank–field surveys were identically disheartening: no emission–line primeval galaxies were found.

The main issue hampering the unsuccessful searches was a general failure to achieve sufficient limiting flux sensitivities. By the mid–1990s, searches for primeval Lyα had achieved flux levels and survey volumes which, according to simple models, ought to have detected thousands of objects (Pritchet 1994), and yet inexplicably, the population remained elusive. The situation was finally remedied by the commissioning of 10–meter class telescopes. By exploiting the substantial gain in depth afforded by these unprecedented apertures (especially when coupled with the growing availability of large–format mosaic CCDs well–suited for wide–field imaging and spectroscopic multiplexing), emission–line searches ultimately did uncover a substantial population of high–redshift Lyα–emitters, albeit with Lyα fluxes nearly two orders of magnitude fainter than originally predicted.

That said, the search techniques used in past decades are fundamentally the same as the techniques used to great success today. In this thesis, we present results from our efforts to detect and characterize primeval galaxies by their signature high–redshift Lyα emission lines utilizing two observational techniques: serendipitous slit spectroscopy and narrowband imaging. By pushing these techniques to their utmost limits, we probe the Lyα–emitting galaxy population out to redshifts as high as $z \approx 6.5$, bordering and perhaps breaching the very epoch of reionization (e.g. Malhotra & Rhoads 2004; Stern et al. 2005). In the currently favored Λ–cosmology[1], with $\Omega_M = 0.3$, $\Omega_\Lambda = 0.7$, and $H_0 = 70$ km s^{-1} Mpc^{-1}, galaxies at this epoch reside in a universe which is just 800 million years old, a mere 6% of its current age. We now motivate this effort by situating it within the field of observational cosmology as a whole (§ 1.1) and by comparing competitive techniques for detecting high–redshift objects (§ 1.2 and § 1.3). We conclude with an overview of the thesis to follow (§ 1.4).

1.1   History: Breaching the Realm of the Nebulae

Modern observational cosmology, with its modest aim of "understanding the entire universe and all its contents" (Peacock 1999), finds its foundation in the interplay between a handful of revelations made during the first few decades of the 20th century. Chief among these discoveries was the confirmation that the spiral nebulae are indeed "island universes" unto themselves (Opik 1922; Hubble 1925, 1926), with masses comparable to that of our own galaxy. Subsequent measurements of the distances to the nebulae, now known to be extra–galactic, yielded the surprising result of Hubble's Law, that the recession speed of galaxies as indicated by their redshifts is linearly related to their distances (Hubble 1929).

The rapid acceptance of this startling observational result was no doubt due in part to the fact that the theoretical underpinnings for an expanding universe had recently come into place. Einstein's General Theory of Relativity (Einstein 1915a,b,c), coupled with the assumption of a homogeneous, isotropic universe, required a world model which (in

---

[1] Unless otherwise specified, we use this cosmology throughout.

the absence of a cosmological constant) either expands or contracts. Friedmann (1922) gave the solution for the expanding case, which was rediscovered by Lemaitre (1927) and subsequently connected to the burgeoning phenomenon of redshifted galaxies. As a result, in the synergy between the new body of cosmology theory and the growing catalog of extra–galactic observations was born the notion that the universe expanded from an initial hot, dense state in which the radiation density dominates the matter density — an idea which was alternately welcomed as a harbinger of new physics (Lemaitre 1931), and vilified for its philosophical repugnance (Eddington 1931).

On either interpretation, this standard cosmological model leaves open the question of how the universe evolved from its smooth initial state to its current state, in which the typical distance between the highly discrete groups of matter known as galaxies is on the order of megaparsecs, while the typical sizes of the galaxies themselves are on the order of kiloparsecs. In fact, that galaxies exist at all as such remarkable departures from homogeneity remains one of the most fundamentally striking cosmological mysteries. Under the current Cold Dark Matter paradigm, structure forms from the growth of very tiny adiabatic density perturbations in the primordial dark matter distribution (e.g. Eggen et al. 1962; Sandage et al. 1970; Peebles 1971; Press & Schechter 1974). The density perturbations grow and collapse under gravitational instability, and the resulting overdensities cause gas to accumulate, cool, and form stars (e.g. White & Rees 1978). Thereafter structure grows hierarchically, with proto–galactic clumps forming at comparatively high redshift ($z \gtrsim 10$), and then merging into larger galaxies and finally clusters (e.g. Baron & White 1987; Baugh et al. 1998).

Observationally, the standard cosmological model has the simple consequence that — because the speed of light is finite — light detected from distant galaxies was emitted when the universe was significantly younger. Thus, to identify and characterize high–redshift galaxies is to look back in time, to observe the ancestors of the present–day galaxies in the very midst of their formation and evolution. The search for primeval galaxies therefore provides direct access to the simplest and most fundamental lines of physical inquiry: how will the world end, how did it begin, and what happened in between.

1.2   The Race to High Redshift

As a consequence, the quest for the most distant galaxy has historically been of fundamental importance to observational cosmology. Prior to the 1950s, the dominant search technique consisted of selecting galaxy clusters by their overdensities on deep photographic plates. This technique, however, is inherently limited. Because clusters require a substantial fraction of the Hubble time to form and virialize, exploiting clusters as high–luminosity landmarks for probing the extra–galactic universe in optical light is useful only to $z \ll 1$. See Stern & Spinrad (1999) for a brief review.

The high–redshift game was soon revolutionized by Baade & Minkowski (1954) with the identification of Cygnus A — one the brightest radio sources in the sky — as a galaxy at $z = 0.056$. This landmark discovery ushered in nearly four decades during which radio–selection proved to be the most effective tool for finding high–redshift objects. Much of this work was performed by Spinrad and collaborators (e.g. Spinrad et al. 1985) by way of spectroscopic follow–up to objects in the Revised Third Cambridge Catalogue (3CR; Bennett 1962), a compendium of the brightest radio sources in the Northern hemisphere. Radio–selection as a high–redshift search technique owed its success to the strong Malmquist bias inherent in radio flux density–limited samples. The 3CR contains the most powerful radio galaxies in the sky, and emission line luminosity and radio power are correlated (e.g. McCarthy 1993). Follow–up to 3CR members therefore selected targets with the most luminous emission lines at a given redshift, a fact that Spinrad et al. exploited to such an extent that they sometimes had a spectroscopic redshift at the radio source position before the host galaxy had been identified in optical imaging.

By the mid–1980s, much of the 3CR had been observed spectroscopically, yielding a then–record of $z = 1.82$ (Spinrad & Djorgovski 1984), and new techniques were developed to push to yet higher redshifts. Merely lowering the limiting flux density of complete radio samples did not prove to be effective, as it allowed for low-redshift, low radio–luminosity interlopers to contaminate the pool of high–redshift candidates. As such, two techniques which add an additional layer of selection were developed. The first technique selected candidates at the peak of the ~ 1 Jy radio source counts whose hosts have the faintest observed 2.2 μm (K–band) fluxes. This sample resulted in the first radio galaxy at $z > 3$ (Lilly 1988). Yet more successful was the technique of selecting sources with ultrasteep radio spectra. This technique was based on the discovery that the fraction of radio source

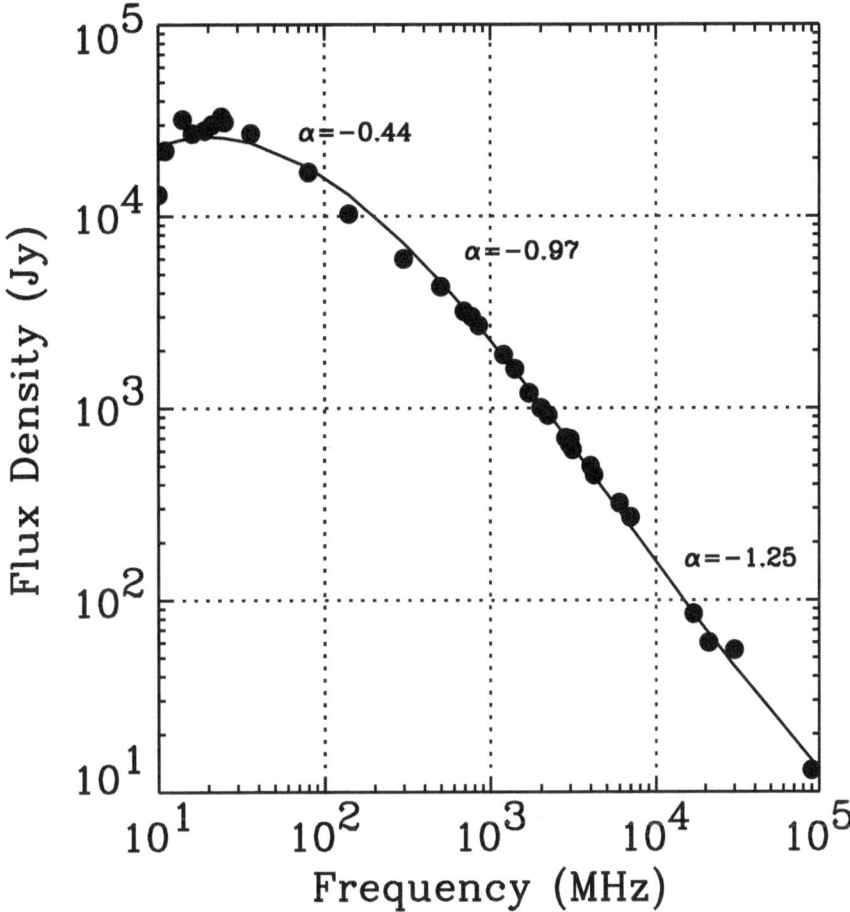

FIG. 1.1.— Radio spectral energy distribution of Cygnus A (adapted from de Breuck 2000). Note that the radio spectra index α, defined by $S_\nu \propto \nu^\alpha$, steepens with increasing frequency. Locating Cygnus A at larger distances would therefore cause an observer to measure increasingly steep spectral indices between any two observed frequencies.

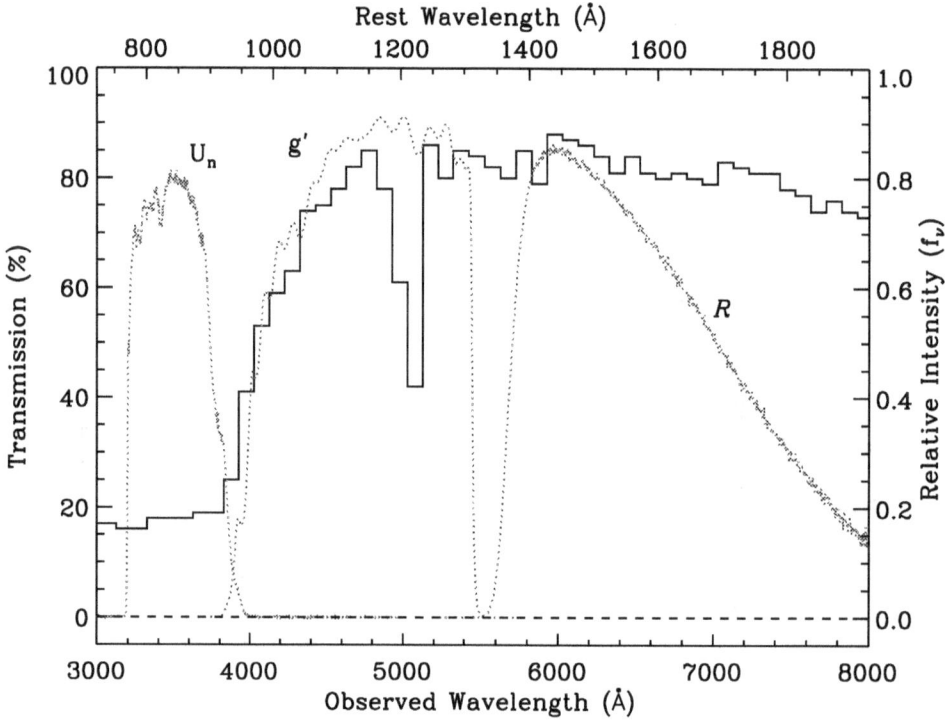

FIG. 1.2.— Model star–forming galaxy at $z = 3.151$, overlaid with three KPNO broadband filters: U_n, g', and R (adapted from Steidel & Hamilton 1992). Note the that the U_n and g' bandpasses are optimally placed to measure the Lyman continuum decrement at $\lambda_{obs} = (1 + z) \times 912$ Å for a star–forming galaxy at $z \sim 3$.

host galaxies identified in the Palomar Observatory Sky Survey plates decreases with steep-ening radio spectra index (Tielens et al. 1979; Blumenthal & Miley 1979), and that the most powerful radio galaxies locally have radio spectral energy distributions which steepen with frequency (Figure 1.1; e.g. Carilli & Yun 1999). Therefore, cosmological k–corrections result in a steepening of radio spectral index with increasing redshift between any given observed frequencies. The selection of ultrasteep spectrum sources resulted in the detec-tion of several radio galaxies with $z > 4$ (Lacy et al. 1994; Rawlings et al. 1996), and in the discovery of a record–setting radio galaxy at $z = 5.19$ (van Breugel et al. 1999).

As of the mid–1990s, radio galaxies remained the only population of galaxies detected beyond $z > 3$. With the advent of the 10–meter class telescopes, however, revolutionary optical selection techniques re–emerged to dethrone the radio methods which had domi-

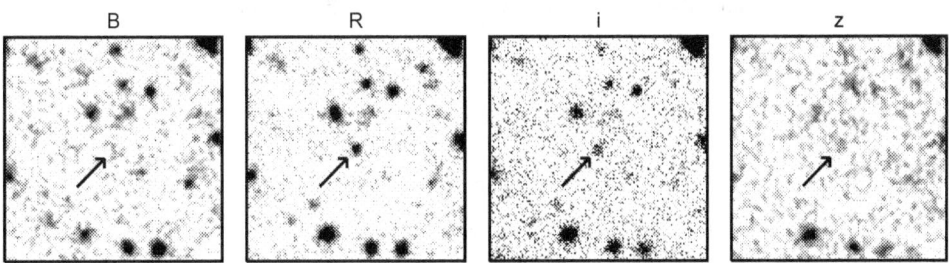

FIG. 1.3.— Example of a $z \approx 4$ B–band "drop–out" galaxy in the Pisces field (Stern 2001, private communication). The imaging is from the Palomar 5m and the KPNO 4m. The images shown are $34''$ on a side; north is up and east is to the left. The source is faintly detected in B ($B \approx 27$), strongly detected in R ($R = 24.8$) and visible in both i ($i \approx 25$) and z ($z \approx 25$). The large discontinuity between B and R is due to absorption of the drop–out galaxy's restframe UV continuum photons by the Lyα and Lyβ forests at $z \leq 4$.

nated for so long. A consequence of this shift in the dominant selection wavelength was a shift in the character of the high–redshift sources selected. Radio galaxies tend to represent rare beasts in the cosmos: they are massive, spatially rare, luminous over many decades of the electromagnetic spectrum, and typically harbor powerful active nuclei or intense bursts of star formation (e.g. McCarthy 1993). Optical selection techniques, on the other hand, generally select a more "normal," less active population of distant galaxies (e.g. Stern & Spinrad 1999).

The cornerstone of modern optical selection was laid with the photometric technique outlined in Steidel & Hamilton (1992), and then implemented by Steidel et al. (1996b) to identify a large population of star–forming field galaxies at $3 \lesssim z \lesssim 3.5$. Photometric selection relies on the fact that the restframe UV spectrum of a star–forming galaxy will be dominated by hot O and B stars, resulting in a flat UV continuum ($f_\nu \propto \nu^0$ for $\lambda > 1216$ Å) and a pronounced continuum discontinuity at the Lyman break (912 Å), as photons more energetic than 912 Å get absorbed in the stellar photospheres of the very stars whose light dominates the spectrum (Figure 1.2). At cosmological redshifts, attenuation due to the Lyα and Lyβ forests make an additional (and generally dominant) contribution to the UV spectral discontinuity, absorbing photons with wavelengths shortward of the Lyα line, at $\lambda_{\rm obs} = (1 + z) \times 1216$ Å. This spectral break can easily be picked off in broadband photometry, with the net effect that high–redshift objects effectively "drop out" out of bluer passbands (Figure 1.3). When such candidates are targeted for follow–up spectroscopy with the current generation of large–format spectroscopic multiplexers, $\sim 10 - 30$ galaxies can be confirmed in a single observation, easily exceeding the *total* number of known high–redshift radio galaxies at comparable redshifts.

Since its overwhelmingly successful application to galaxies at $z \sim 3$, the next obvious application of photometric selection was to push to higher redshifts by selecting drop–out galaxies in successively longer–wavelength passbands. Particularly when combined with the exceptionally deep, spaced–based imaging provided by the Hubble Deep Field (HDF, Williams et al. 1996), or more recently by the Great Observatories Origins Deep Survey (GOODS, Giavalisco et al. 2004a), photometric selection has yielded scores of star–forming galaxies first to $z > 4$ (e.g. Dickinson 1998; Steidel et al. 1999), then to $z > 5$ (e.g. Spinrad et al. 1998; Weymann et al. 1998; Iwata et al. 2003; Bunker et al. 2003; Dickinson et al. 2004), and currently to $z > 6$ and beyond (e.g. Bunker et al. 2004; Bouwens et al. 2004; Stanway et al. 2003, 2004a,b; Nagao et al. 2004). At comparatively brighter magnitudes, $i' - z'$ colors from the Sloan Digital Sky Survey have been used to identify QSOs out to $z = 6.4$ (Fan et al. 2003). These galaxies and quasars are broaching the very edge of the epoch of cosmic reionization (Becker et al. 2001; Djorgovski et al. 2001). Photometric selection has therefore opened the door to the very dark ages of the universe.

1.3 Emission–Line Selection

While photometric selection was successfully exploiting the continuum properties of high–redshift star–forming galaxies, a handful of optical selection methods arose which take advantage of their expected emission–line properties. Nascent galaxies may be thought of as cosmological H II regions which have been photo–ionized in their first bursts of star formation, and consequently ought to have spectral energy distributions dominated by strong emission lines. To wit, stellar population synthesis models predict Lyα restframe equivalent widths of $50 - 200$ Å for young, dust–free galaxies (e.g. Charlot & Fall 1993). By $z > 3$, Lyα is observable in the optical passband, and under standard conditions (i.e. case B recombination theory, the assumption of an optically thick nebula), approximately two Lyα photons are produced for every three ionizing photons emitted by the stars. Thus, Lyα emission is a natural choice to serve as the observational signpost of primeval galaxies in formation (Partridge & Peebles 1967). Notably, selecting on Lyα emission is

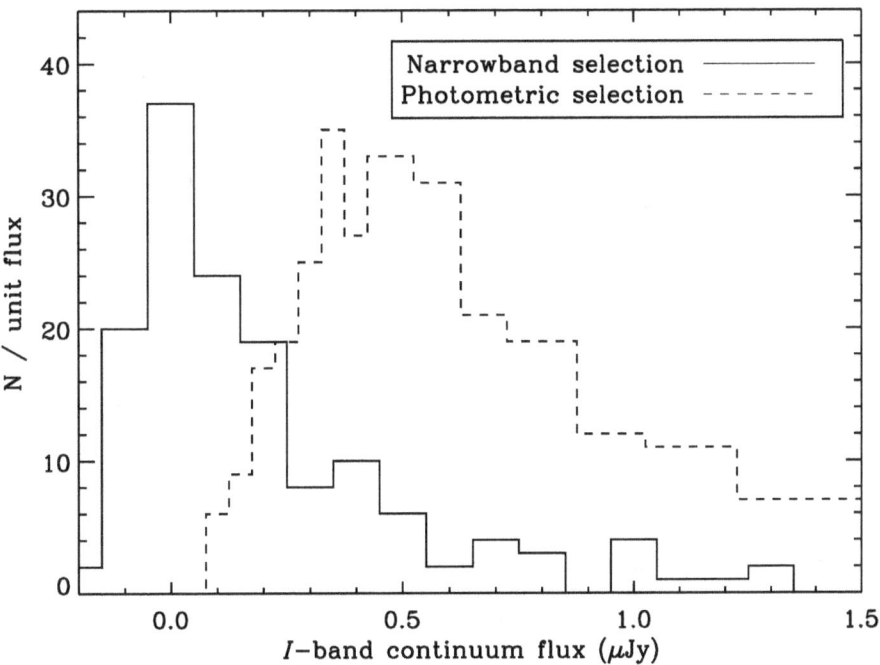

FIG. 1.4.— Distribution of *I*–band continuum flux for narrowband–selected Lyα–emitting galaxies in the Large Area Lyman Alpha survey (LALA; solid line; $<z> = 4.5$) and for photometrically–selected Lyman–break galaxies in the NOAO Deep Wide–Field Survey (NDWFS; dotted line; $<z> = 3.6$) (Dey 2004, private communication). Narrowband–selection detects star–forming galaxies with fainter continua than those accessible to photometric–selection, but only selects that fraction of sources with strong line emission. The LALA and NDWFS surveys are discussed in greater detail in Chapters 5 and 6. Note, 0.5 µJy in *I*–band flux corresponds to an AB magnitude of $I_{AB} \sim 24.5$.

complementary to selecting on the Lyman break (Figure 1.4). The Lyman break method at typical sensitivities requires moderately bright galaxies, while Lyα searches can identify intrinsically fainter galaxies. Though it will only select the fraction of objects that have strong line emission, Lyα–selection probes further down the continuum luminosity function than Lyman–break selection alone can achieve.

Chief among the techniques for selecting high–redshift Lyα emission are serendipitous slit spectroscopy and narrowband imaging, though integral–field spectroscopy, narrowband

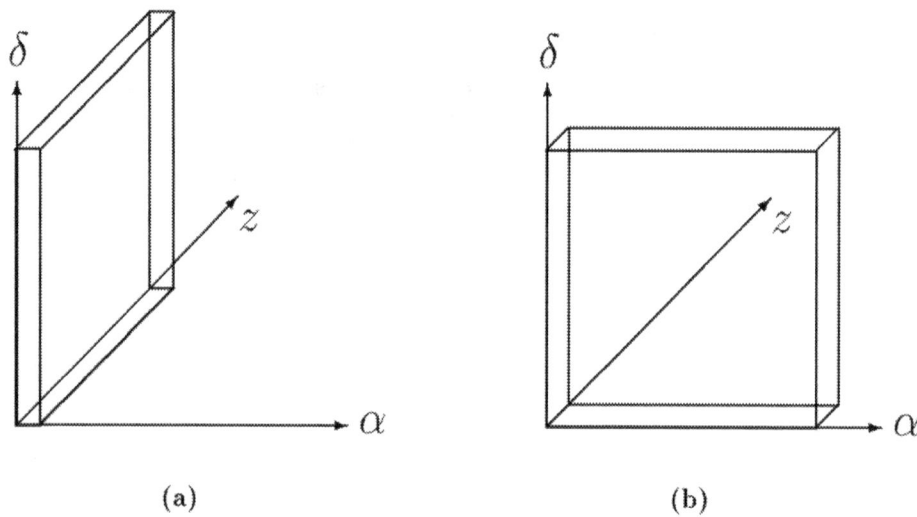

(a)   (b)

FIG.   1.5.— Survey volumes for complementary emission–line search strategies: (a) serendipitous slit spectroscopy, and (b) narrowband imaging.

spectroscopy, and slitless spectroscopy have also shown some success (e.g. van Breukelen et al. 2005; Crampton & et al. 2000; Koo & Kron 1980, respectively). Each method offers a variety of strengths and weakness. As we discuss below, serendipitous slit spectroscopy and narrowband imaging are each capable of achieving useful flux limits ($\lesssim 10^{-17}$ erg s^{-1} cm^{-2}) in reasonable integration times, but slit spectroscopy is generally restricted to small survey volumes, and narrowband imaging to comparatively narrow slices in redshift–space (Figure 1.5). Slitless spectroscopy allows a large volume to be probed, but because each object is superimposed on a night–sky background from all wavelengths passed by the optics, long integration times are required to achieve acceptable flux limits. An integral–field unit samples an area on the sky in which each pixel contains a spectrum, covering both a large volume and a large redshift range, but to date no such survey has achieved the combination of volume and sensitivity necessary to be competitive with serendipitous slit spectroscopy or narrowband imaging at the highest accessible redshifts ($z > 4.5$).

Clearly, no single technique is perfectly suited to reliably sample the global population of Lyα–emitters and to trace their evolution over cosmological timescales. The remainder of this thesis will focus on serendipitous slit spectroscopy and narrowband imaging, which are highly complementary, and which have enjoyed by far the greatest successes of any of the emission–line search methods. We now briefly elaborate on each in turn.

1.3.1   Serendipitous Slit Spectroscopy

Even blind spectroscopic observations, provided they are sufficiently deep, are sensitive to line–emitting sources which serendipitously fall within the slit. Moreover, spectroscopy is an efficient way to achieve excellent limiting fluxes for detecting unresolved emission lines, in part because the higher spectral resolution helps lower contamination by the sky background in any given resolution element. In a 1.5 hour spectrum at moderate resolution ($\lambda/\Delta\lambda \approx 1000$) with the Low Resolution Imaging Spectrometer (LRIS) on the Keck I and II telescopes, the limiting flux probed for a spectroscopically unresolved emission line in a $1''$ aperture is better than $S_{\mathrm{lim}}(5\sigma) \approx 10^{-17}$ erg s^{-1} cm^{-2}, though this limit is strongly wavelength–dependent (Stern & Spinrad 1999).

Obviously, the major limitation to this method is that the survey volume is fairly restricted. The redshift sensitivity to Lyα spans fully $3 \lesssim z \lesssim 6.5$; the limits are set by the plummeting response of optical CCDs in the UV and the near–IR, as well as by the strong OH and O_2 night–sky emission lines which dominate ground–based observations at wavelengths longer than 9300 Å. However, the spectroscopic slit generally offers an exceptionally small solid angle. The angular size of a typical longslit on Keck/LRIS is $175'' \times 1.0''$, corresponding to a comoving volume of ~ 600 Mpc^3, more than two orders of magnitude less than the survey volumes available to contemporary narrowband imaging surveys.

Even still, the small survey volume offered by slit spectroscopy is offset by the unparalleled efficiency of relying on serendipity. *Any* reasonably deep program of slit spectroscopy simultaneously doubles as a serendipitous survey; therefore, serendipitous surveys require no direct allocation of telescope time. And the long list of unprecedented discoveries made serendipitously in extra–galactic astronomy alone is a testament to its viability as a search method. Serendipitous detections include the discovery of a galaxy cluster at $z = 2.40$ (Pascarelle et al. 1996), at least three quasars at $z > 4$ (McCarthy et al. 1988; Schneider et al. 1994, 2000), many galaxies at $z > 3$ (e.g. Manning et al. 2000), and the first object detected at $z > 5$ (Dey et al. 1998); recently, serendipitous spectroscopy breached $z > 6$ (Stern et al. 2005).

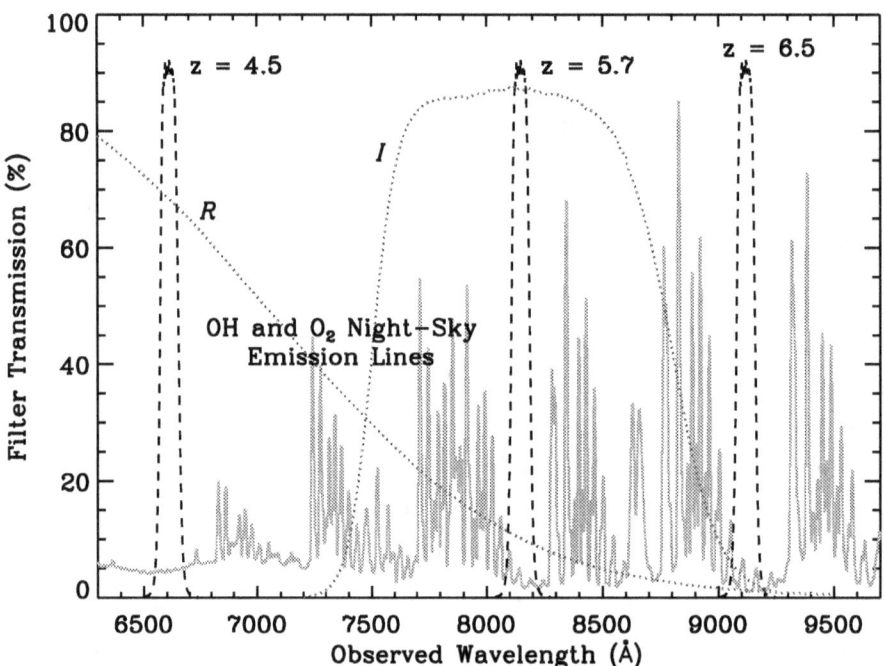

FIG. 1.6.— Night–sky emission spectrum overlaid with sample narrowband and broadband comparison filters typical of searches for high–redshift Lyα, in this case, the Large Area Lyman Alpha (LALA) survey. The filters depicted attempt to minimize contamination by the sky background by targeting windows in night–sky emission. As such, they are sensitive to Lyα at $z \approx$ 4.5, 5.7, and 6.5, sampling star–forming galaxies in a universe which is ∼ 10%, 7%, and 6% of its current age, respectively.

1.3.2   Narrowband Imaging

In direct contrast to serendipitous slit spectroscopy, narrowband imaging surveys impose a restriction on the redshift range covered, but they offer a much larger solid angle. With a modern 8k × 8k mosaic CCD and a narrowband filter with FWHM ∼ 80 Å, a single narrowband image probes a search volume of ∼ 10^5 Mpc³. By targeting windows in the night–sky emission lines (Figure 1.6), narrowband imaging can achieve limiting flux densities comparable to those of slit spectroscopy with only a factor of ∼ 3 increase in the integration time. Selecting high–redshift Lyα in narrowband imaging exploits the fact that an emission line which lies in the narrowband filter will easily stand out against the continuum detected in a comparison broadband filter. This technique can be combined with shorter–wavelength "veto" filters to eliminate foreground galaxies by their continuum signal blueward of the emission line; high–redshift Lyα–emitters have no such signal, owing to continuum blanketing by the Lyα and Lyβ forests. An additional layer of selection can be added in the form of an equivalent width requirement. Large continuum–selected (Cowie et al. 1996; Hogg et al. 1998) and Hα–selected (Gallego et al. 1996) samples indicate that restframe equivalent widths of [O II] $\lambda3727$ — a common low–redshift doppleganger of narrowband–selected Lyα — are typically $W_\lambda^{\mathrm{rest}} \ll 100$ Å, though at least one exceptional source with $W_\lambda^{\mathrm{rest}} \approx 600$ Å at $z = 1.464$ is known (Stern et al. 2000b). Nonetheless, narrowband selection generally requires follow–up spectroscopy to verify the identity of the emission line, making it less efficient than spectroscopy alone.

Narrowband imaging has been the dominant technique for systematically detecting high–redshift Lyα–emitters since its first successful "blank–sky" implementation in the late 1990s (Cowie & Hu 1998; Hu et al. 1998). Narrowband surveys are now routinely employed to systematically assemble samples of Lyα–emitting galaxies at redshifts $z \approx 4.5$ (Malhotra & Rhoads 2002), $z \approx 5.7$ (Rhoads & Malhotra 2001; Rhoads et al. 2003; Hu et al. 2004), and $z \approx 6.5$ (Kodaira et al. 2003; Rhoads et al. 2004; Taniguchi et al. 2005).

1.4   Thesis Overview

In this thesis, we report on our efforts to detect and characterize primeval galaxies by their signature high–redshift Lyα emission lines. In Part I, we describe the results of our serendipitous slit spectroscopy survey of the Hubble Deep Field and its environs. Chapter 2 contains our moderate–resolution ($\lambda/\Delta\lambda = 250 - 375$) efforts with Keck/LRIS, which resulted in a catalog of 74 spectroscopic redshifts spanning $0.10 < z < 5.77$, including a galaxy cluster at $z = 0.85$ and five galaxies at $z > 5$. In Chapter 3, we focus on a single high–redshift Lyα–emitter (ES1), serendipitously detected in echelle spectroscopy at $z = 5.190$. At the time of its discovery, ES1 was one of only nine known galaxies at $z > 5$, and was the sixth most distant known galaxy. The unprecedented spectral purity of the observation offered evidence for a galaxy–scale outflow with a velocity of $v > 300$ km s^{-1}, consistent with wind speeds observed in powerful local starbursts (typically 10^2 to 10^3 km s^{-1}), and with simulations of the late–stage evolution of Lyα emission in star–

forming systems. Finally, in Chapter 4 we present our observations of the remarkable source CXOHDFN J123635.6+621424, which is both the highest redshift known spiral galaxy, and a rare example of a high redshift, hard X–ray–emitting Type II AGN. Significantly, each of these results was acquired with no direct allocation of telescope time.

We turn to our narrowband imaging result in Part II. Chapters 5 and 6 present a catalog of 76 $z \approx 4.5$ Lyα–emitting galaxies spectroscopically–confirmed in campaigns of Keck/LRIS and Keck/DEIMOS follow–up observations to candidates selected in the Large Area Lyman Alpha (LALA) narrowband imaging survey. The resulting sample of confirmed Lyα emission lines show large equivalent widths (median $W_\lambda^{\mathrm{rest}} \approx 80$ Å) but narrow physical widths ($\Delta v < 500$ km s^{-1}), supporting the conclusion of Malhotra et al. (2003) and Wang et al. (2004) that the Lyα emission in these sources derives from star formation, not from AGN activity. Moreover, though the expectation from theoretical models of galaxy formation in the primordial Universe is that a small fraction of Lyα– emitting galaxies at $z \approx 4.5$ may be nascent, metal–free objects ("Population III," e.g. Scannapieco et al. 2003), and indeed we found with 90% confidence that 3 to 5 of the confirmed sources exceed the maximum Lyα equivalent width predicted for normal stellar populations, we did not detect He II λ1640 emission in either individual or composite spectra. That is, though these galaxies are young, they show no evidence of being truly primitive, Population III objects. Finally, we construct a luminosity function of $z \approx 4.5$ Lyα emission lines for comparison to Lyα luminosity functions created from similar surveys spanning $3.1 < z < 6.6$. We find that if there is evolution in the Lyα luminosity function between $z \approx 3$ and $z \approx 6$, its significance is below the statistical uncertainty of these data. We summarize and give thoughts about future directions in Chapter 7.

Part I

Serendipitous Slit Spectroscopy

Chapter 2

Serendipitously Detected Galaxies in the Hubble Deep Field

A version of this chapter was previously published in *The Astronomical Journal* (Dawson, S., Stern, D., Bunker, A. J., Spinrad, H., & Dey, A. 2001, AJ, 122, 598). Reproduced by permission of the AAS.

Abstract

We present a catalog of 74 galaxies detected serendipitously during a campaign of spectroscopic observations of the Hubble Deep Field North (HDF) and its environs. Among the identified objects are five candidate Lyα–emitters at $z \gtrsim 5$, a galaxy cluster at $z = 0.85$, and a Chandra source with a heretofore undetermined redshift of $z = 2.011$. We report redshifts for 25 galaxies in the central HDF, 13 of which had no prior published spectroscopic redshift. Of the remaining 49 galaxies, 30 are located in the single–orbit HDF Flanking Fields. We discuss the redshift distribution of the serendipitous sample, which contains galaxies in the range $0.10 < z < 5.77$ with a median redshift of $z = 0.85$, and we present strong evidence for redshift clustering. By comparing our spectroscopic redshifts to optical/IR photometric studies of the HDF, we find that photometric redshifts are in most cases capable of producing reasonable predictions of galaxy redshifts. Finally, we estimate the line–of–sight velocity dispersion and the corresponding mass and expected X–ray luminosity of the galaxy cluster, we present strong arguments for interpreting the Chandra source as an obscured AGN, and we discuss in detail the spectrum of one of the candidate $z \gtrsim 5$ Lyα–emitters.

2.1 Introduction

The Hubble Deep Field North (Williams et al. 1996, hereafter W96) ranks among the most thoroughly studied portions of the extragalactic universe. The extremely deep multi–color images obtained with the WFPC2 camera on the *Hubble Space Telescope*, reaching AB mag $B_{450} \sim 29$ with $0\rlap{.}''1$ resolution, have revolutionized our understanding of the faint galaxy population and have yielded diverse new results in observational cosmology. Follow–up observations to the original survey span the electromagnetic spectrum, from the radio (Fomalont et al. 1997; Richards et al. 1998; Richards 2000) to the sub–millimeter (Hughes et al. 1998; Barger et al. 2000), to both ground and space–based near infrared (Hogg et al. 1997; Dickinson 1997, 2000; Thompson & Storrie–Lombardi 1997) and far infrared (Aussel et al. 1999). Recently, X–ray data have become available (Hornschemeier et al. 2000, 2001), and UV observations with the Space Telescope Imaging Spectrograph are in progress (Gardner et al. 2000). In addition to imaging, numerous groups are pursuing spectroscopic observations of galaxies in the HDF. Cohen et al. (2000) report on a magnitude–limited sample more than 92% complete to Vega mag $R = 24$; Steidel et al. (1996a) and Lowenthal et al. (1997) report on color–selected samples of Lyman–break galaxies at $z \sim 3$; while Zepf et al. (1997) report on a morphologically selected sample of probable gravitational lenses. See Ferguson et al. (2000) for a review of measurements and phenomenology of sources in the HDF across the electromagnetic spectrum.

Consequently, the HDF and the eight adjacent, single–orbit I_{814} Flanking Fields (see W96, Table 2) now constitute a very rich database for the study of galaxy formation and evolution. Early results included the confirmation of a flattening in the slope of the faint elliptical/S0 galaxy number count–magnitude relation (Abraham et al. 1996; Zepf 1997), as well as the revealed inadequacy of the Hubble sequence as a classification scheme for galaxies fainter than $I_{AB} > 24$ mag (Abraham et al. 1996). The selection of four very broad bandpass filters for the WFPC2 observations was driven partly by the desire to identify high–redshift galaxies via the Lyman–break technique. Indeed, this strategy facilitated the discovery of distant galaxies whose Lyman–breaks have been redshifted into the U–band (Steidel et al. 1996a; Lowenthal et al. 1997), the B–band (Steidel et al. 1999), and beyond (Spinrad et al. 1998; Weymann et al. 1998). The exquisite resolution of the

WFPC2 images spurred considerable effort toward quantifying galaxy morphology, leading to the disentanglement of morphological k–correction from morphological evolution, and revealing an increase in the fraction of true irregulars at faint magnitudes/high redshift (Bunker et al. 1999). Most recently, mining of this data–rich field has yielded refined techniques in estimating photometric redshifts (e.g. Fernández–Soto et al. 1999) and has produced dramatic implications for the history of star–formation (Madau et al. 1996; Steidel et al. 1999) as well as for the role of dust in the distant universe (Hughes et al. 1998; Ouchi et al. 1999).

We are pursuing a variety of programs to study distant galaxies in the HDF. The primary science from these observations, discussed elsewhere, includes extremely deep ($\sim 10^{\mathrm{h}}$) moderate and high–resolution Keck/LRIS spectroscopy of Lyman–break galaxies at $z \sim 3$ aimed at understanding their stellar populations and galactic dynamics (e.g. Bunker et al. 1998), and low–resolution spectroscopy of B–band and V–band dropouts whose colors suggest a population of galaxies with Lyman–breaks and significant Lyα– forest absorption at $z > 4$ (Spinrad et al. 1998). In the course of these observations, we have targeted more than 65 galaxies in the HDF and its environs for deep spectroscopy, and in so doing we have serendipitously observed some 125 objects which were located propitiously along the slit of a target. Out of the sample of serendipitous detections, we have determined redshifts for 74 sources, with 25 galaxies in the HDF proper, 30 galaxies in the HDF Flanking Fields, and 19 galaxies beyond but in the vicinity of the Flanking Fields. Thirteen of the detections in the central HDF provided the first ever spectroscopic redshift determinations for those sources.

From the first detection of pulsars to the discovery of the cosmic microwave back-ground, serendipity has historically made significant contributions to astronomy. In extra–galactic astronomy in particular, dramatic serendipitous detections include the discovery of a galaxy cluster at $z = 2.40$ (Pascarelle et al. 1996), at least three quasars at $z > 4$ (McCarthy et al. 1988; Schneider et al. 1994, 2000), and the discovery of the first object at $z > 5$ (Dey et al. 1998). Serendipity plays a less dramatic but still significant role in large scale redshift surveys: serendipitous detections make up roughly 8% of the mea-sured galaxies in the complete $K_s < 20$ mag galaxy sample presented by Cohen et al. (1999). Serendipitous surveys in their own right are efficient, as they require no direct initial allocation of telescope time, and they have proven to be both competitive with and complementary to narrow–band imaging surveys. See Thompson & Djorgovski (1995),

Manning et al. (2000), and Stern et al. (2000a) for reports on serendipitous searches for high–redshift Lyα emission.

Though none of the serendipitous detections reported herein constitute singularly momentous discoveries, given the status of the HDF as ranking among the most thoroughly mapped pieces of the extragalactic universe, we would be remiss not to report all galaxy redshifts determined in the course of our observations of this well–studied field. In §2.2 we discuss the spectroscopic observations and the data reduction. In §2.3 we describe the redshift determination and the process by which the serendipitously detected galaxies were visually identified. We present the catalog of serendipitously detected galaxies in §2.4, and we discuss their distribution in redshift space, the comparison between spectroscopic and photometric redshifts, the observed properties of the galaxy cluster at $z = 0.85$, the observed properties of the Chandra source at $z = 2.011$, and the candidate $z \gtrsim 5$ Lyα–emitters in §2.5. Throughout this paper, we adopt an Einstein–de Sitter cosmology with $H_0 = 100 \, h_{100}$ km s^{-1} Mpc^{-1}, $q_0 = 0.5$, and $\Lambda = 0$. All quoted magnitudes are in the AB system[1] unless otherwise specified.

2.2   Observations and Data Reductions

Between 1997 February and 2001 February, we obtained deep spectra of photometrically selected high–redshift candidates in the HDF and its environs. The data were taken with the Low Resolution Imaging Spectrometer (LRIS; Oke et al. 1995) at the Cassegrain focus on the 10m Keck I and Keck II telescopes. The camera uses a Tek 2048^2 CCD detector with a pixel scale of $0\farcs212$ pixel^{-1}. To maximize observing efficiency, we exclusively used the dual amplifier readout mode. Except for three longslit observations in 1997 February, the data were taken with slitmasks designed to obtain spectra for ~ 15 targets simultaneously.

For the vast majority of observations, we used a 150 lines mm^{-1} grating blazed at 7500 Å, which produces a 4.8 Å pix^{-1} dispersion. The spectral coverage with this grating is approximately 4000 Å to 1 micron, allowing us observe the entire optical window irrespective of the grating tilt or the location of the slit on the slitmask. We used a 300 lines mm^{-1} grating blazed at 5000 Å (2.55 Å pix^{-1} dispersion) on one set of observations, a

---

[1] The AB magnitude system is defined such that $m(AB) = -2.5 \log(f_\nu) - 48.60$ with f_ν measured in erg s^{-1} cm^{-2} Hz^{-1} (Oke & Gunn 1983). The value of the constant is set by the condition $m(AB) = V$ for a flat–spectrum source.

400 lines mm^{-1} grating blazed at 8500 Å (1.86 Å pix^{-1} dispersion) on two sets of observations, and a 600 lines mm^{-1} grating blazed at 5000 Å (1.28 Å pix^{-1} dispersion) on one set
of observations. For targets within the central HDF (where the astrometric solutions are
well–determined), we employed 1″0 slits, yielding a spectral resolution of $\lambda/\Delta\lambda_{\mathrm{FWHM}} = 375$
with the 150 lines mm^{-1} grating. For targets outside of the HDF (where the astrometric
solutions are less well–determined), we employed 1″5 slits, yielding a spectral resolution of
$\lambda/\Delta\lambda_{\mathrm{FWHM}} = 250$ with the 150 lines mm^{-1} grating. The minimum set of exposures for any
given target was 3 × 1800s. As the position angle of the slit for a particular target normally
changed from observation to observation, only a small number (~ 5) of the serendipitous
detections benefited from re–observation.

The most recent set of observations (2001 February) were made with the advent of
the LRIS–B spectrograph channel (McCarthy et al. 1998). For these observations, we used
the 400 lines mm^{-1} grating blazed at 8500 Å in red channel, and a 300 lines mm^{-1} grism
blazed at 5000 Å (2.64 Å pix^{-1} dispersion) in the blue channel. To split the red and blue
channels, we used a dichroic with a cutoff at 6800 Å. Together, the two channels in this
set–up afforded a spectral coverage of roughly 3200 Å to 1 micron. Again, we observed
the entire optical window, but at almost twice the dispersion of our typical spectrograph
configuration.

We used the IRAF[2] package (Tody 1993) to process both the longslit and the slitmask data, following standard slit spectroscopy procedures. Some aspects of treating the
slitmask data were facilitated by a home–grown software package, BOGUS[3], created by
D. Stern, A.J. Bunker, and S.A. Stanford. Wavelength calibrations were performed in the
standard fashion using Hg, Ne, Ar, and Kr arc lamps; we employed telluric sky lines to adjust the wavelength zero–point. We performed flux calibrations with longslit observations
of standard stars from Massey & Gronwall (1990) taken with the instrument in the same
configuration as the relevant multislit observation. However, it should be noted that the
absolute scale of the fluxed spectra must be regarded with caution. Not all of the nights
were photometric; there may be substantial slit losses in the case of an extended source;
small errors in slitmask alignment cause additional light loss; and since the position angle
of an observation was set by the desire to maximize the number of targets on a mask, the

---

[2]IRAF is distributed by the National Optical Astronomy Observatories, which are operated by the
Association of Universities for Research in Astronomy, Inc., under cooperative agreement with the National
Science Foundation.

[3]BOGUS is available online at http://zwolfkinder.jpl.nasa.gov/~stern/homepage/bogus.html.

observations were in general not made at or near parallactic angle. Moreover, in the case of serendipitous detections, it is unlikely that the object is optimally aligned with the slit even when all other parameters are perfect. Fortunately, it is merely the redshift of a given object — not the absolute flux or continuum shape — which is of interest at present.

2.3   Visual Identifications and Redshift Determinations

2.3.1   Visual Identifications

A serendipitous detection in spectroscopic data presents two challenges to the observer: (1) to locate on an image of the field the progenitor of the spectroscopic signature, and (2) to determine the nature of the object and, where possible, its redshift. We now address the former problem; we discuss the latter in the following section.

In some respects, the process of cataloguing serendipitous detections proceeds backwards from the usual steps involved in compiling redshifts. Generally, one begins with photometry for a galaxy whose location is known and subsequently obtains a spectrum. In our case, we began with a spectrum and worked backward to the progenitor's location and photometry. To accomplish this task, we combined what was known about the observation — the location of the target, the dimensions and orientation of the target slit, and the position of the target *within* the slit — and we reconstructed the position of the slit on the sky. We then mapped the reconstructed slit image to the target field and thereby identified *a posteriori* the objects which we in fact observed.

In the most favorable cases, the two–dimensional spectra contained multiple serendipitous detections. By comparing the relative spatial separations between continuum detections in the two–dimensional spectra to the separations between sources on the slit image, we could uniquely identify each of the progenitors. Even under unfavorable circumstance, in which the target was too faint to appear in the spectrum or was mis–aligned with the slit, the progenitor of a lone serendipitous detection could be identified by comparing its position on the two–dimensional spectrum to its position in the image relative to the edges of the slit.

To this end, the galaxies in the sample divide into two categories: those within the central HDF, and those without. For galaxies inside of the central HDF, we made the visual identification by mapping the slit image to the remarkably deep, well–resolved central I_{814} images presented in W96. We label the galaxies in Table 2.2 by their IDs, isophotal magnitudes, and positions as given in that paper. If, on the other hand, the target slit extended outside of the central HDF, we relied on supporting photometry provided by the single–orbit I_{814} Flanking Field images of W96, the deep Hawaii 2.2m V, I, $H+K$ images of Barger et al. (1999, hereafter B99), or the deep U_n, G, R images of Steidel et al. (1996c). Since there is no existing nomenclature for sources beyond the central HDF, we adopted a labeling scheme in Table 2.3 based on the galaxy positions.

To facilitate this position–based nomenclature for serendipitous observations of galaxies in the Flanking Fields and beyond, we computed an astrometric solution to the Hawaii 2.2m I–band image of B99. From a fit to 72 objects in a $10' \times 10'$ portion of the digitized POSS–II plates[4] (obtained via the Digitized Sky Survey[5]), we found a platescale of $0\rlap{.}''189$ pixel^{-1} and an orientation angle of $0.630°$, both of which are consistent with the values reported by B99. The dispersion in the fit was $0\rlap{.}''22$ in the right ascension (RA) direction and $0\rlap{.}''26$ in the declination (Dec) direction, giving a total error of $0\rlap{.}''37$. This error is comparable to the error reported by B99. As a check to the fit, we compared the RA and Dec positions of 10 objects in the central HDF as given by W96 against our newly computed Hawaii 2.2m I–band positions and found a mean absolute offset of $0\rlap{.}''11$ in RA and $0\rlap{.}''16$ in Dec, for a total mean offset of $0\rlap{.}''20$. This error is smaller than the sum in quadrature of the total dispersion in our fit and the accuracy of the W96 absolute astrometry (reported as good to approximately $0\rlap{.}''4$), suggesting that our reported RA and Dec positions themselves are good to roughly $0\rlap{.}''4$.

In three cases, the serendipitous detection fell outside of the Hawaii 2.2m fields. To visually identify these objects, we utilized our own 70 minute R–band image taken with the Echelle Spectrograph and Imager (ESI, Sheinis et al. 2000) on UT 2000 May 05. See Stern et al. (2000b) for a detailed account of the observation and data reduction. The astrometric solution for the position–based nomenclature was determined exactly as described for the Hawaii 2.2m image; I_{AB} magnitudes are not available for these detections. In five cases, the progenitor of a serendipitous spectroscopic detection was too faint to be detected in any of the supporting imaging. Nonetheless, we were able to estimate a position for the

---

[4] The Second Palomar Observatory Sky Survey (POSS-II) was made by the California Institute of Technology with funds from the National Science Foundation, the National Aeronautics and Space Administration, the National Geographic Society, the Sloan Foundation, the Samuel Oschin Foundation, and the Eastman Kodak Corporation.

[5] The Digitized Sky Survey was produced at the Space Telescope Science Institute under U.S. Government grant NAG W-2166. The images of these surveys are based on photographic data obtained using the Oschin Schmidt Telescope on Palomar Mountain and the UK Schmidt Telescope. The plates were processed into the present compressed digital form with the permission of these institutions.

source by extrapolating from the known position of the target and the dimensions and orientation of the target slit. We have indicated such cases on Table 2.3.

2.3.2   Redshifts

For each member of the serendipitous catalog, we measured the redshift by visually inspecting the spectrum and noting the wavelengths of spectral features. For objects with multiple strong emission lines, the proper interpretation of the spectral features was unambiguous and the assignment of their rest wavelengths was straightforward. The situation was more difficult for faint objects showing only absorption lines. If such a spectrum did not conform to the standard pattern of Balmer lines and the H+K Ca II doublet seen in the vicinity of the 4000 Å–break (D4000), then it was generally impossible to determine a redshift.

The most common type of serendipitous detection involved the presence of a single emission line, the interpretation of which can problematic. In general, a single, isolated line could be any one of Lyα, [O II], Hβ, [O III] λ5007, or Hα, though given sufficient spectral coverage, most erroneous interpretations can be ruled out. For instance, the absence of Hβ or [O III] λ4959 serves to discount the interpretation of a solo line as [O III] λ5007. Similarly, lines that are bluer than rest Hα cannot be Hα themselves, and the presence of Hβ or [O III] λ5007 would be expected for a solo line redder than rest Hα (but see Stockton & Ridgway 1998). Hence, the primary threat to determining one–line redshifts is the potential for mis–identifying Lyα as [O II] or vice versa. Unfortunately, with low dispersion spectra it is often impossible to distinguish between the high equivalent width forms of these emission lines without a pronounced continuum depression or a line asymmetry, both characteristic of Lyα. For two accounts of the potential pitfalls associated with one–line spectroscopic redshift identifications, see Stern et al. (2000a) and Manning et al. (2000).

In part to reflect the uncertainty in interpreting solo lines, we divide the serendipitous detections into five spectral categories (SC) based on their general morphology. Table 2.1 lists the spectral categories with a brief description of each. The spectra of category 1 sources show multiple features which can be uniquely identified, yielding secure redshift determinations. The spectra of category 2 sources show a solo emission line in the presence of strong continuum both redward and blueward of the line. Such lines were identified as [O II], and the redshift determination is considered secure. The spectra of category 3 sources show a solo emission line redward of a strong continuum break. Such lines were identified as Lyα and the continuum breaks were interpreted as the onset of absorption by the Lyα–forest (which causes significantly diminished flux shortward of 1216 Å). Of course, especially in star–forming systems, the continuum in the vicinity [O II] can also show a break — the Balmer break at 4000 Å — and in cases of low signal–to–noise, the morphology of the Balmer break alone is not sufficient to distinguish it from the break at Lyα. Fortunately, for galaxies at $z \gtrsim 4$, the break at Lyα is expected to be of greater amplitude than the largest observed D4000 amplitudes (see Stern & Spinrad 1999, Fig. 12), so the two features can be easily discerned. At lower redshifts, however, the amplitude of the two breaks may be comparable, and without corroborating spectral features the redshift identification is largely subjective. Of five category 3 sources in this catalog, two are at $z \gtrsim 4$, one has a redshift which is confirmed by other authors, and one has supporting photometric redshifts from two other authors; their redshift determinations are considered secure. The redshift of the remaining category 3 source should be considered tentative, as indicated on Table 2.3. Example spectra for categories 1, 2, and 3 are shown in Figure 2.1.

The spectra of category 4 objects show an isolated emission in the absence of any continuum, which generally suggests a weak detection of either [O II] or Lyα. Clearly, the confidence one can exercise in discriminating between these two cases is a strong function of the robustness of the detection, the resolution of the spectrum, and the availability of supporting imaging. See §2.5.5 for a detailed discussion of the redshift determination of a typical category 4 source. Example spectra for both interpretations of category 4 sources are shown in Figure 2.2. The spectra of category 5 sources show a continuum break. Such breaks were classified as either the Balmer break or as Lyα–forest absorption according to the strength of the continuum blueward of the break. Example spectra for both interpretations of category 5 are shown in Figure 2.3. The redshift determinations of both category 4 and category 5 sources are considered secure unless otherwise indicated. Serendipitous detections about which we were unable to attain a reasonable degree of confidence were omitted from the catalog; nearly half of the initial sample of 121 serendipitous detections were rejected for this reason.

To minimize the possibility that we mis–classified the solo emission line of a low–redshift category 4 source as high–redshift Lyα, we checked that the source as visually identified in the Hawaii 2.2m I–band image of B99 did not also appear in the R–band image of Steidel et al. (1996c). In this fashion, we ensured that the R–band flux of the

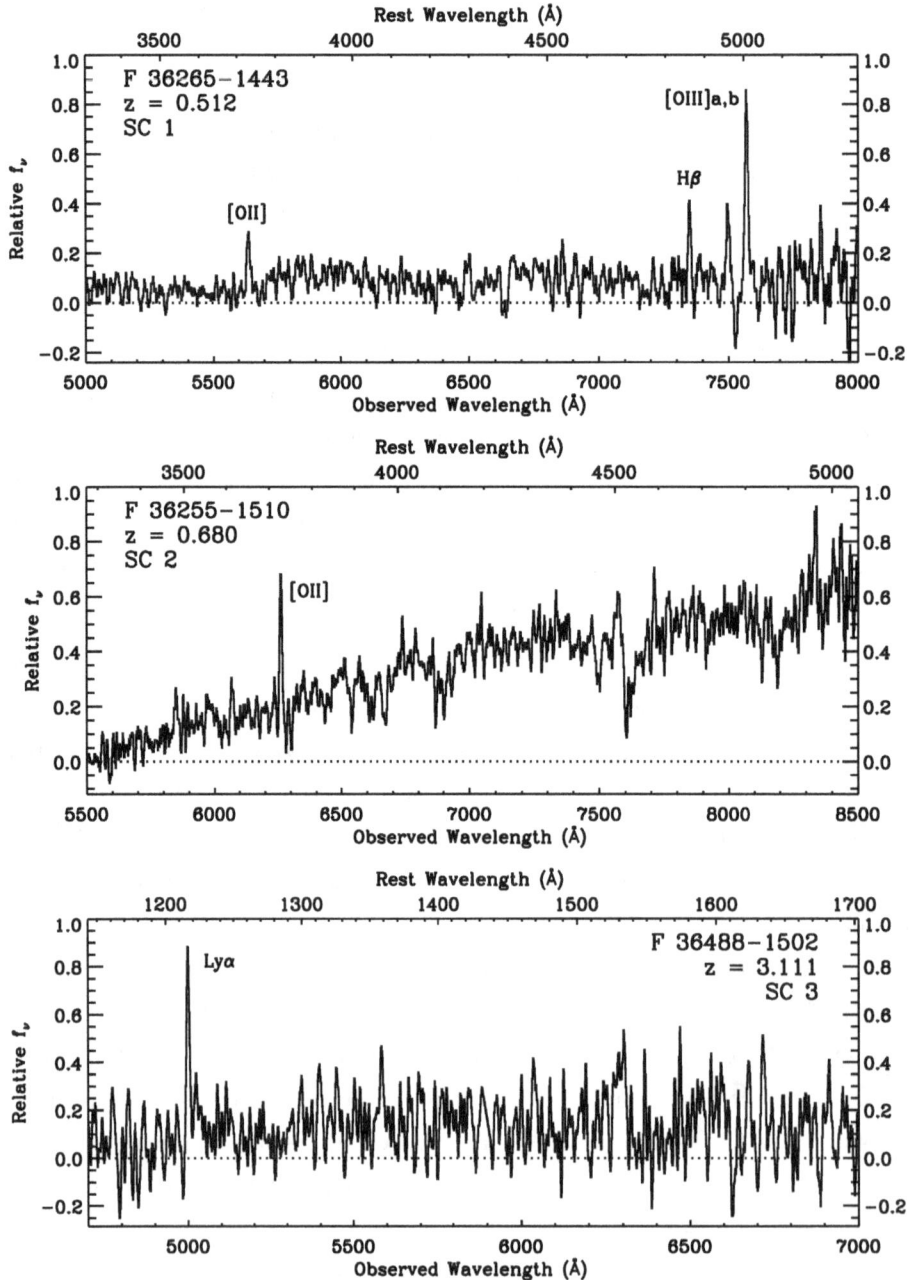

FIG. 2.1.— Example spectra for spectral categories 1, 2, and 3. See §2.3.2 and Table 2.1 for a detailed account of the spectral categories. The total exposure time for each is 5.4 ks. The spectra have been smoothed using a 20 Å boxcar filter.

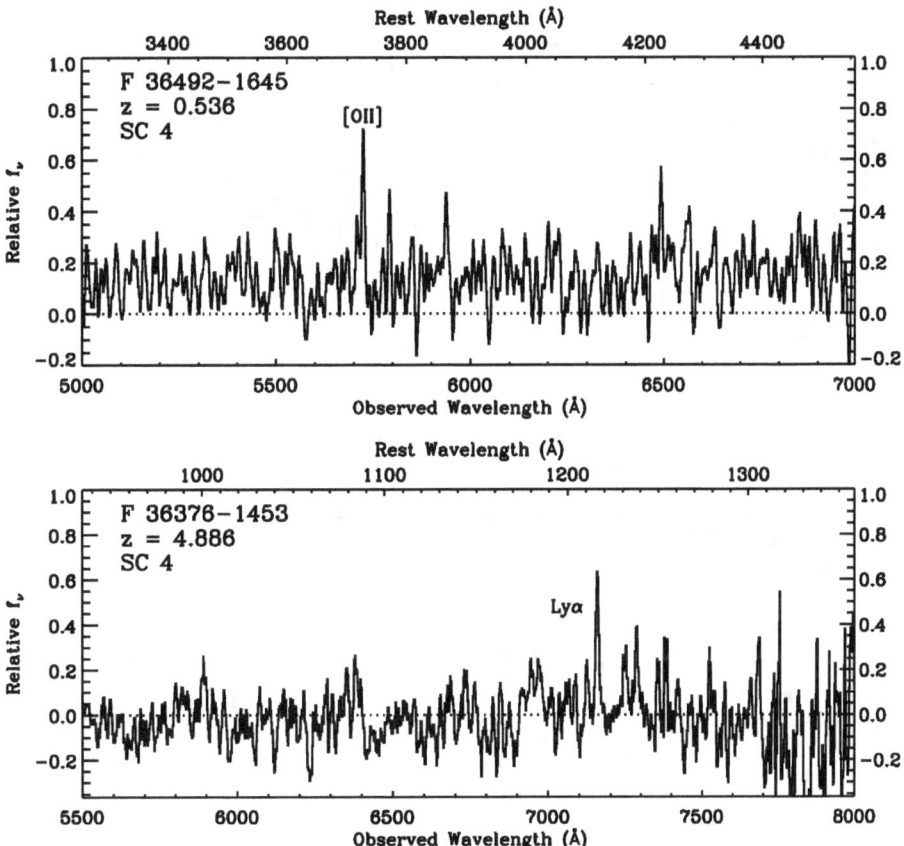

FIG. 2.2.— Example spectra for spectral category 4, in which a solo emission line in the absence of continuum is identified as either [O II] (top panel) or as Lyα (bottom panel), based in part on the line profile, the line observed–frame equivalent width, and/or the supporting imaging. See §2.3.2 and Table 2.1 for a detailed account of the spectral categories. The total exposure time for each is 5.4 ks. The spurious features observed in the continuum are due to residuals from the subtraction of strong telluric OH emission lines. The blueward "shoulder" on the [O II] emission line is an imperfectly removed cosmic ray. The spectra have been smoothed using a 20 Å boxcar filter.

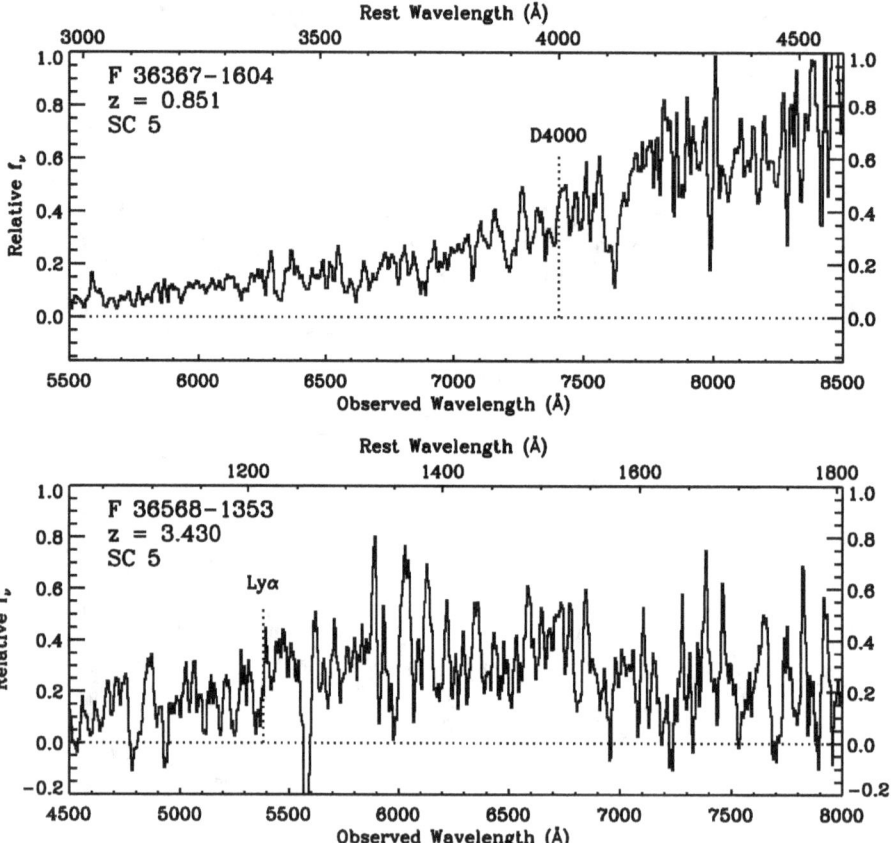

FIG. 2.3.— Example spectra for spectral category 5, in which a continuum break is interpreted as the 4000 Å–break (top panel) or as the onset of Lyα–forest absorption (bottom panel) according the strength of the continuum blueward of the break. See §2.3.2 and Table 2.1 for a detailed account of the spectral categories. The total exposure time for each is 5.4 ks. The spurious features observed in the continuum are due to residuals from the subtraction of strong telluric OH emission lines. The spectra have been smoothed using a 20 Å boxcar filter.

source in question was severely attenuated by the hydrogen forest, consistent with $z \gtrsim 4$. We discovered one erroneous redshift determination with this technique: F 36265–1443 was marginally detected in 1999 June such that [O III] $\lambda5007$ appeared in the two–dimensional spectrum as a solo emission line at $\lambda = 8136$ Å, and the line was initially mis–classified as high–redshift Lyα at $z = 5.691$. However, the presence of the progenitor in the R–band image of Steidel et al. (1996c) ruled out the tantalizing high–redshift interpretation, and subsequent targeted spectroscopy revealed a spectrum with [O II], [O III] $\lambda4959$, [O III] $\lambda5007$, Hβ, and Hγ in emission at $z = 0.625$.

In the event that a redshift for a serendipitous detection remained undetermined, one possibility is that the object lies in the so–called "redshift desert," the interval spanning roughly $1.7 < z < 2.3$. The limits of this interval are set by the fact that at higher redshifts Lyα would fall on the detector, and at lower redshifts the oxygen lines and/or the Balmer lines would fall on on the detector. At intermediate redshifts, however, there is a dearth of prominent spectral features, rendering redshift determination difficult. A second possibility is that the object does have spectral features which are in principle observable, except that the features fall in a region heavily contaminated by night sky emission. As sky subtraction is particularly problematic at $\lambda > 7200$ Å for low signal–to–noise, low dispersion spectra, it is reasonable to conclude that at least a few redshifts were lost to this effect.

It should be noted that for the ~ 5 cases in which a single galaxy was multiply observed, the agreement in the individual redshifts was excellent. Discrepancies never exceeded $\Delta z = 0.004$.

2.4   The Catalogs

We present the catalog of serendipitously detected galaxies in Table 2.2 and Table 2.3. Table 2.2 contains 25 galaxies located in the HDF proper, identified by their W96 number as described in the preceding section. The I_{814} magnitude is the isophotal magnitude given by W96, and the RA and Dec are J2000 coordinates, also given therein. The spectral category was assigned as described in §3.1; also see Table 2.1. Table 2.3 contains 49 galaxies located outside the central HDF, identified by their positions as described in the preceding section. The 30 galaxies located in the HDF Flanking Fields are indicated. The isophotal I_{AB} magnitudes were determined by running the source extraction algorithm SExtractor (Bertin & Arnouts 1996) on the Hawaii 2.2m I–band image of B99. We estimated the I_{AB}

zero–point by using stars in the central HDF; as such, the uncertainty in the I_{AB} is ~ 0.3 mag. All spectral lines in both tables are emission lines unless otherwise noted.

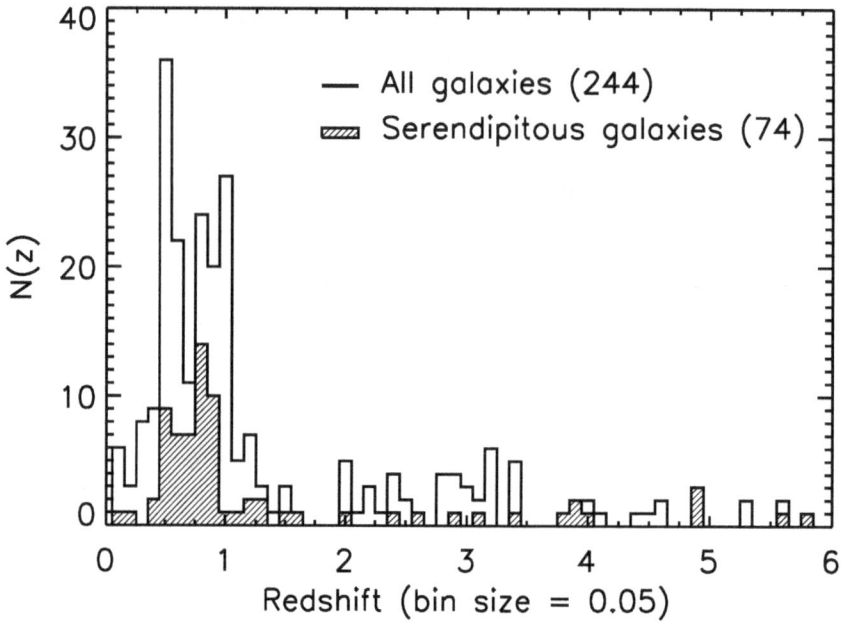

FIG. 2.4.— Distribution of redshifts of the serendipitous sample compared to a total sample consisting of all published redshifts for galaxies in the central HDF plus 26 galaxies which flank the central HDF.

2.5   Discussion

The 74 galaxies in the serendipitous catalog span the redshift range $0.10 < z < 5.77$, with a median redshift of $z = 0.85$. The vast majority of the galaxies are emission-line systems; 5% of the sample show only absorption lines. This bias stems from the diminished likelihood of serendipitously detecting an absorption–line system with sufficient signal–to–noise to allow the redshift to be determined.

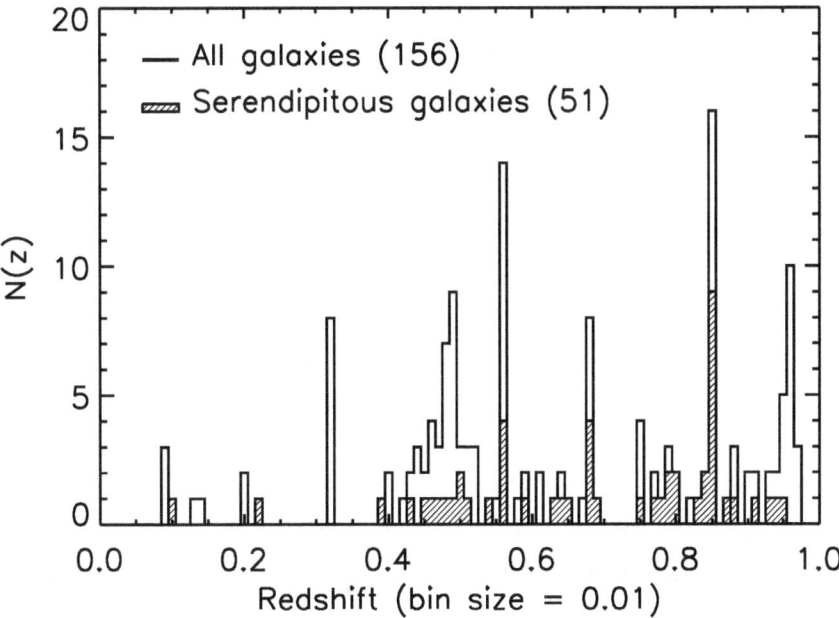

FIG. 2.5.— Distribution of redshifts of the serendipitous sample compared to the total sample for the range $0 < z < 1$.

2.5.1 The Redshift Distribution

The redshift distribution of the serendipitous catalog, compared with a "total sample" consisting of this sample, all published redshifts for galaxies in the central HDF, and 26 published redshifts for galaxies flanking the central HDF, is shown in Figures 2.4 and 2.5. Sources for the total sample are Bunker et al. (1998); Cohen et al. (1996); Cohen et al. (2000); Lowenthal et al. (1997); Phillips et al. (1997); Spinrad et al. (1998); Stern & Spinrad (1999); Waddington et al. (1999); and Weymann et al. (1998). The histogram displayed in Figure 2.4 displays the total range of redshifts of the combined catalogs, $0.089 < z < 5.77$, with a comparatively coarse resolution of $\Delta z = 0.1$. Given the caveat that we are insensitive to galaxies in the redshift range $1.7 < z < 2.3$ (cf. §3.1), we find that the redshift distribution of the serendipitous sample closely follows that of the total sample.

To investigate the redshift clustering properties of the serendipitous sample, we display the redshift distribution for the galaxies in the range $0 < z < 1$ with a resolution of $\Delta z = 0.01$ in Figure 2.5. The figure shows clear evidence of clustering in both the serendipitous sample and the total sample. Moreover, the clustering present in the total sample is mirrored almost perfectly by that present in the serendipitous sample. Assuming a fixed number of galaxies per redshift bin (i.e. no evolution in bin membership with redshift), we find a 2.3σ peak in the serendipitous sample at $z = 0.79$, a 3.2σ peak at $z = 0.56$ and $z = 0.68$, and a 6.9σ peak at $z = 0.85$. In total, we find that 17 out of the 51 serendipitous galaxies (33%) fall into peaks significant at greater than 97.5% confidence. This figure compares favorably with that of Cohen et al. (1996, hereafter C96), who find that 57 out of 140 (41%) of their sample of spectroscopically observed HDF galaxies fall into redshift peaks. That the locations of our peaks vary somewhat from those in C96 is not surprising. Whereas C96 chose redshift bins of variable centers and widths so as to maximize their significance relative to occurring by chance in a smoothed velocity distribution, we chose fixed bin centers and widths, cf. Phillips et al. (1997). Even so, our peaks centered on $z = 0.56$ and $z = 0.68$ no doubt reflect the same structures revealed by the peaks in C96 at $z_p = 0.559$ and $z_p = 0.680$, respectively. We find no evidence of periodicity in the peak redshifts, as described by Broadhurst et al. (1990).

Beyond the strong evidence of redshift clustering, there are two outstanding features of the redshift distribution of the serendipitous sample. First, there is a relative deficiency of serendipitous detections at $z < 0.4$. Second, the redshift peak centered on $z = 0.32$ evident in the total sample is not represented in the serendipitous sample. Taken together, these features appear to suggest a selection effect which excludes galaxies at $z < 0.4$ from serendipitous detection. However, since this redshift range is perfectly accessible to LRIS via the Balmer lines and by [O II] and [O III] emission, it is likely that the scarcity of low–redshift galaxies in the serendipitous catalog is merely the combined effect of: (1) the increasingly small cosmological volume surveyed at low–redshift, (2) the comparatively small size of the serendipitous catalog, and (3) the fact that the HDF was selected to be devoid of bright galaxies in the first place. At a minimum, these facts make it impossible to comment on the significance of the apparent $z < 0.4$ deficiency.

2.5.2   A Check of Photometric Redshifts

Photometric redshift techniques have become an essential tool of observational cosmology, with applications ranging from determining luminosity functions to selecting high–

redshift candidates for spectroscopy. We have utilized our set of spectroscopic redshifts for 23 of the 25 serendipitously detected galaxies in the central HDF to carry out a test of the photometric redshifts presented by Fernández–Soto et al. (1999), who employ a maximum-likelihood analysis applied to spectral energy distribution–fitting of precise U_{300}, B_{450}, V_{606}, I_{814}, J (1.2 μm), H (1.65 μm), and K (2.2 μm) photometry. For two galaxies, HDF 4–402.1 and HDF 4–236.0, no photometric redshift was available, no doubt owing to their faintness: I_{AB} = 24.96 and 28.26, respectively. The sample of predicted redshifts was taken from the group's world wide web site — the University of New South Wales/State University of New York at Stony Brook HDF Clickable Map[6] — which is an interactive version of the catalog presented in the associated paper.

We compare the spectroscopic redshift (z_{spec}) and the photometric redshift (z_{phot}) in a scatter plot of z_{spec} versus z_{phot} for redshifts less than 1.5 in Figure 2.6. There are three obvious errors in the photometric redshifts: (1) HDF 4–639.1, listed with z_{spec} = 2.592 and z_{phot} = 0.000, whose spectrum shows Lyα in emission with a strong continuum break (SC 3), and whose z_{spec} is confirmed by both Steidel et al. (1996a) and Cohen et al. (2000); (2) HDF 2–600.0, listed with z_{spec} = 0.425 and z_{phot} = 1.800, whose spectrum shows a strong solo emission line interpreted as [O II] (SC 4); and (3) HDF 4–658.0, listed with z_{spec} = 0.558 and z_{phot} = 4.320, whose spectrum shows both [O II] and [O III] emission (SC 1). These outliers comprise 13% of the sample, roughly consistent with the finding of Cohen et al. (2000) that outliers at more than 4σ in the z_{spec}–z_{phot} plane comprised $\sim$ 10% of the subset of galaxies at $z < 1.5$. The outliers are not shown in Figure 2.6, as they are off the scale.

The mean and the dispersion of the difference between the predicted photometric redshifts and the measured spectroscopic redshifts are $|\Delta z|$ = 0.380 and $\sigma_{\Delta z}$ = 0.907, respectively. However, these values are dominated by the three discrepant points described above. When the discrepant points are omitted, we find a mean deviation of $|\Delta z|$ = 0.098 and a dispersion of $\sigma_{\Delta z}$ = 0.010. These errors are consistent with the assessment that cosmic variance (the fact that the model spectra used in determining photometric redshifts represent a finite sample of all possible galaxy spectra) rather than photometric errors is the dominant source of error at small redshift (Fernández–Soto et al. 1999). Moreover, these results confirm that — barring catastrophic errors — photometric redshifts are capable of producing reasonable predictions of galaxy redshifts where suitably precise multicolor

---

[6] http://bat.phys.unsw.edu.au/~fsoto/hdf/hdf_fs.html

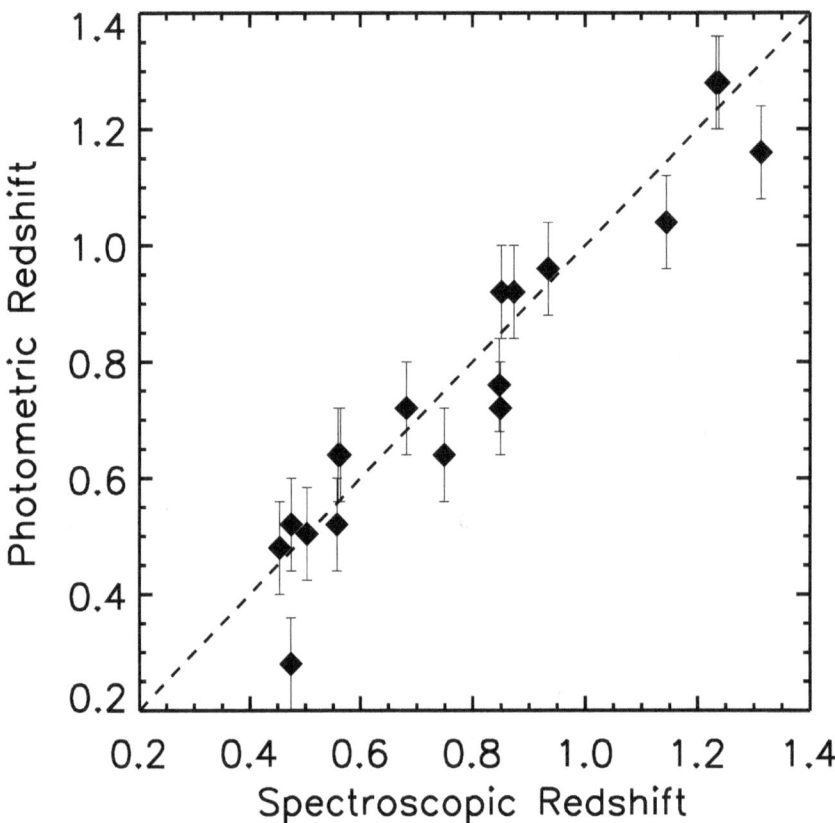

FIG. 2.6.— Comparison of spectroscopic and photometric redshifts for 17 serendipitously detected galaxies in the central HDF. The error bars are attributable to cosmic variance, the fact that the model spectra used in determining photometric redshifts represent a finite sample of all possible galaxy spectra; photometric errors are negligible in this redshift range. Three obviously erroneous photometric redshifts are off the scale: HDF 4–639.1, listed with $z_{\mathrm{spec}} = 2.592$ and $z_{\mathrm{phot}} = 0.000$; HDF 2–600.0, listed with $z_{\mathrm{spec}} = 0.425$ and $z_{\mathrm{phot}} = 1.800$; and HDF 4–658.0, listed with $z_{\mathrm{spec}} = 0.558$ and $z_{\mathrm{phot}} = 4.320$.

photometry is available.

2.5.3   A Galaxy Cluster at z = 0.85

We report the serendipitous discovery of ClG 1236+6215, a galaxy cluster with redshift $z = 0.85$ nominally centered at $\alpha = 12^h 36^m 39\overset{s}{.}6$, $\delta = +62°15'54''$ (J2000). The cluster was initially identified as an over–density of centrally concentrated red objects in a small region to the northwest of the HDF in the deep Hawaii 2.2m V and I images of Barger et al. (1999). In a circle of radius 45 arcsec centered on the cluster position, the density of objects with $(V - I)_{AB} > 1.5$ is 18 arcmin^{-2}, versus a density of only 6.5 arcmin^{-2} over the rest of the 90 arcmin^2 Hawaii 2.2m field. We interpreted the $(V - I)_{AB}$ color of the concentration to be the result of the 4000 Å break redshifted into the I–band, and we targeted five of the reddest members for spectroscopy. All five of the targets proved to have redshifts very near to $z = 0.85$. We added three more redshifts by selecting objects from the redshift catalog of Cohen et al. (2000) which had $(V - I)_{AB} > 1.5$ and $0.84 < z < 0.86$, and which were located within 45 arcsec ($0.17\,h_{100}^{-1}$ Mpc) of the cluster center. Together, the eight spectroscopic members of ClG 1236+6215 yield a mean redshift for the cluster of $z = 0.849 \pm 0.004$. The properties of the spectroscopic members of ClG 1236+6215 are summarized in Table 2.4.

Following the prescription of Harrison (1974) for properly considering the contributions to measured redshifts due to the radial component of the motion of our Galaxy with respect to the Local Group, to the cosmological expansion between comoving observers at our Galaxy and at the galaxy cluster, and to the radial component of the peculiar velocity of the galaxy within the cluster, we calculated an estimate of the corrected line–of–sight velocity dispersion $\sigma_\parallel$ in ClG 1236+6215. We followed the treatment of Danese et al. (1980) to account for the spurious systematic contribution to $\sigma_\parallel$ from measurement errors in the member redshifts. Assuming an underlying Gaussian distribution for the galaxy velocities, we found $\sigma_\parallel = 610 \pm 190$ km s^{-1} (68% confidence); this value should be treated with caution due to the small number of spectroscopic members. Beers et al. (1990) point out that the classical standard deviation estimator for cluster velocity dispersions is neither resistant to the presence of outliers nor robust for non–Gaussian underlying populations. However, employing the "gapper" method as implemented in their ROSTAT package yields a correction which is less than our estimated uncertainty.

In the limiting isothermal model, the calculated velocity dispersion translates to a mean cluster mass within a 45 arcsec ($0.17\, h_{100}^{-1}$ Mpc) radius of the cluster center of $M(r < 0.17\, h_{100}^{-1}$ Mpc$) = 2.6 \times 10^{13}\, h_{100}^{-1}\, M_\odot$. For comparison with other authors, the mean mass within the Abell radius is $M(r < 1.5\, h_{100}^{-1}$ Mpc$) = 2.3 \times 10^{14}\, h_{100}^{-1}\, M_\odot$. Of perhaps more immediate observational consequence is the X–ray luminosity expected for the given velocity dispersion. Drawing on a sample of 197 galaxy clusters — which constitutes the largest cluster data set used to date for such a study — Xue & Wu (2000) find $L_X/(10^{43}$ erg s$^{-1}) = 10^{-13.68\pm0.61}\sigma_\parallel{}^{5.30\pm0.21}$ for the X–ray bolometric luminosity–velocity dispersion relation. This result yields an expected X–ray bolometric luminosity for ClG 1236+6215 of $L_X = 1.2 \times 10^{44}$ erg s^{-1}, a value which exceeds the expected detection threshold of the upcoming ≈ 1 Ms *Chandra X–ray Observatory* (CXO) exposure of the HDF and its environs (Brandt et al. 2001).

2.5.4   Spectroscopy of the X–ray Source CXOHDFN J123635.6+621424

Optical spectroscopy of faint X–ray sources is the key to determining the poorly understood physical properties of the population responsible for producing the X–ray background. We present the first published optical spectrum and redshift for CXOHDFN J123635.6+621424, a well–observed X–ray source identified with a face–on spiral galaxy at $z = 2.011$, fortuitously located in the Inner West HDF Flanking Field.

CXOHDFN J123635.6+621424 was first detected as a weak radio source (8.15 μJy at 8.5 GHz; 87.8 μJy at 1.4 GHz) in the sensitive HDF radio surveys of Richards et al. (1998); Richards (2000). The source has a comparatively steep radio spectral index ($S_\nu \propto \nu^{-\alpha}$; $\alpha_{1.4\,\mathrm{GHz}}^{8.4\,\mathrm{GHz}} > 0.87$), and the radio emission extends across 2$''$8. In general, microjansky radio emission from disk galaxies can result from either star formation (e.g. from free–free emission originating in H II regions) or from AGN activity connected with a central engine. Richards et al. (1998); Richards (2000) argued that (1) in the case of a central AGN powering a weak ($P < 10^{25}$ W Hz^{-1}) radio source, the bulk of the radio emission is confined to the nuclear region and is therefore characterized by sub–arcsecond angular scales, and (2) such small scales result in a high opacity to synchrotron self–absorption, yielding flat or inverted spectral indices typically in the range $-0.5 < \alpha < 0.5$. Hence, the origin of the radio emission in CXOHDFN J123635.6+621424 was taken to be extended star–forming regions. This conclusion was ostensibly borne out by an *Infrared Space Observatory* Camera (ISOCAM) detection of the source (Aussel et al. 1999). If the source were a moderate–to–low redshift starburst galaxy (as suggested by Hornschemeier 2001, owing to the object's spatial extent), the ISOCAM 15 μm filter (LW3) would sample rest wavelengths from roughly 6 μm to 12 μm; the mid–infrared emission could therefore be plausibly attributed to the unidentified infrared bands (UIB) and to the hot, 200 K dust which typically dominates the spectral energy distribution of starbursts over those wavelengths (Aussel et al. 1999).

In contradistinction to the foregoing conclusions, both the optical and X–ray properties of CXOHDFN J123635.6+621424 indicate the presence of AGN activity. The optical spectrum shows moderate–width ($\sim$ 1000 km/s), high–ionization emission lines, similar to those of the recently reported quasar II in the Chandra Deep Field South (Norman et al. 2002) and typical of high–redshift radio galaxies (cf. McCarthy 1993; Stern & Spinrad 1999). We detect Lyα, N V λ1240, Si IV λ1397, C IV λ1549, He II λ1640, C III] λ1909, [Ne IV] λ2424, and Mg II λ2800 (Figure 2.7). Moreover, the rest frame equivalent widths of the C III] λ1908 and C IV λ1548 emission lines ($\sim$ 13 Å and $\sim$ 100 Å, respectively) are within the ranges found in multiple AGN emission line surveys and optical/radio quasar surveys (see Lehmann et al. 2000, and references therein). We also note that the C IV λ1549/He II λ1640 ratio of $\sim$ 8 is more typical of quasars than of radio galaxies. Optical and near–IR photometry of CXOHDFN J123635.6+621424 corroborates these findings. Hogg et al. (2000) give $(R - K_s) = 4.74$ for the source, and Hasinger (1999) report that all X–ray counterparts with $(R - K') > 4.5$ in their ROSAT Ultra Deep HRI Survey are either members of high redshift clusters or are obscured AGN. Finally, CXO observations of the source indicate a comparatively hard X–ray spectrum — the definitive signature of an AGN. The X–ray band ratio, defined as the ratio of hard–band (2 keV to 8 keV) to soft–band (0.5 keV to 2 keV) number counts, is $0.75^{+0.71}_{-0.43}$, corresponding to a photon index[7] of $\Gamma = 0.75$ (Hornschemeier et al. 2001).

When re–interpreted in the light of the spectroscopic redshift, even the mid–IR data for CXOHDFN J123635.6+621424 actually indicate the presence of an AGN. For the derived redshift of $z = 2.011$, the ISOCAM LW3 filter samples rest wavelengths spanning only 4 μm to 5 μm. Here, the contribution to the mid–IR spectral energy distribution made by UIB emission and by dust at 200 K is severely attenuated (see Aussel et al. 1999, Figure 1). Hence, the ISOCAM detection of this source is far more plausibly explained by the hot, $\sim$ 10^3 K dust found in the central region of an AGN (e.g. see Aussel et al. 1998) rather than by star formation alone. The weakness of Lyα in this galaxy substantiates the

---

[7]The photon index Γ is derived from a power law model for the X–ray spectrum: $N = AE^{-\Gamma}$, where N is the number of photons s^{-1} cm^{-2} keV^{-1} and A is a normalization constant.

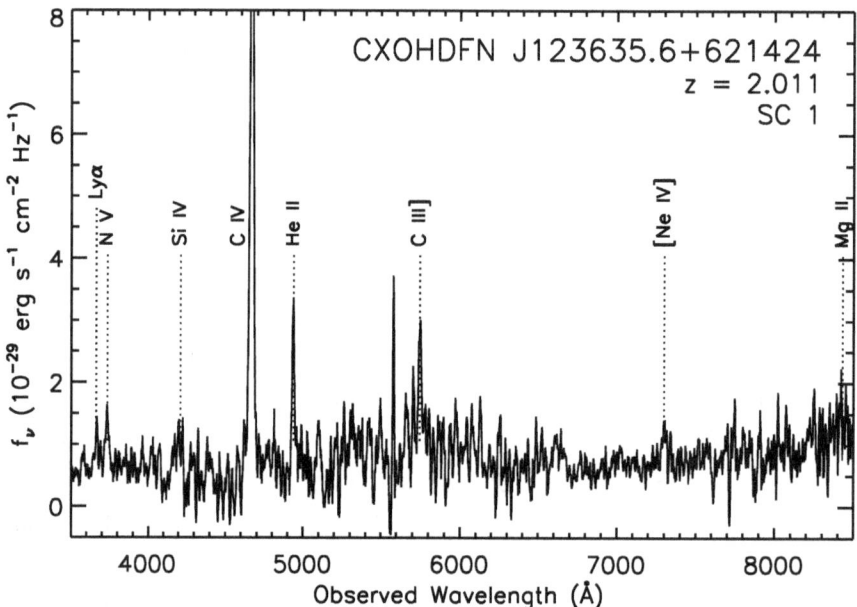

FIG. 2.7.— Optical spectrum of the X–ray source CXOHDFN J123635.6+621424. The spectrum was obtained on UT 24 February 2001, after the advent of the LRIS–B spectrograph channel. Flatfield and flux–calibration difficulties associated with the blue channel prevented us from calibrating the blue side ($\lambda < 6800$ Å) in the standard fashion. To create the spectrum shown, we assumed a flat rest UV spectrum ($f_\nu \propto \nu^0$) and then forced the blue channel and red channel fluxes to agree at 6800 Å. Though line ratios determined within either spectrograph channel (e.g. the $\lambda 1549$/He II $\lambda 1640$ ratio of ~ 8 cited in §2.5.4) are reliable, line ratios made *across* spectrograph channels should be considered suspect. The total exposure time is 8.4 ks. The spectrum was smoothed using a ~ 10 Å boxcar filter.

presence of dust in this system.

Though the canonical wisdom regarding extended radio sources with spectral indices steeper than $\alpha > 0.5$ dictates that such sources are driven by starbursts (Richards et al. 1998; Richards 2000; Hornschemeier et al. 2001), the combined weight of evidence from X–ray, optical, and near– and mid–IR observations of CXOHDFN J123635.6+621424 is definitively in favor of an obscured AGN. This conclusion is consistent with the trend reported by Hornschemeier et al. (2001): that the high X–ray luminosities and large band ratios of several CXO–detected radio sources previously reported as starburst–type objects strongly suggests the presence of heretofore unidentified AGNs. We are currently pursuing Keck/NIRSPEC spectroscopy of this interesting source in order to further detail its physical properties.

2.5.5   Galaxies at z > 5

In the course of deep, targeted spectroscopy of photometric high–redshift galaxy candidates, we have identified several serendipitous high–redshift Lyα–emitting candidates, including five sources at $z \gtrsim 5$. These high–redshift sources are evident in Figure 2.4, and they are listed in Table 2.3. The surface density of such sources is sufficiently high that these discoveries are not unexpected (e.g. Dey et al. 1998; Manning et al. 2000). Indeed, slit spectroscopy surveys for high–redshift Lyα emission are fully complimentary to narrow–band searches (e.g. Hu et al. 1998; Steidel et al. 1999; Rhoads et al. 1999): rather than probing a large area of sky for objects over a limited range of redshift, deep slit spectroscopy surveys a small area of sky for objects over a large range in redshift (Pritchet 1994; Thompson & Djorgovski 1995). The total area covered by the spectroscopic slits during the course of our study was ≈ 2.2 arcmin^2, implying a surface density of ≈ 2.3 arcmin^{-2} Lyα–emitters at redshift $z \sim 5$. This value is roughly consistent with the surface density of high–redshift Lyα–emitters reported by Cowie & Hu (1998): ≈ 3 arcmin^{-2} (unit–z)$^{-1}$ at redshift $z \sim 3.4$, for comparable sensitivity to line flux. Of course, one should exercise caution regarding these values, owing to the small number of detections involved.

Each of the high–redshift sources in this catalog are solo emission line sources (SC 4), and as indicated by a handful of cautionary tales (§3.2 herein; also see Stockton & Ridgway 1998; Stern et al. 2000a), such redshift identifications should be greeted with a degree of circumspection. A detailed discussion of each individual source is beyond the scope of this paper, and a separate manuscript is planned. For now, we restrict the discussion to one

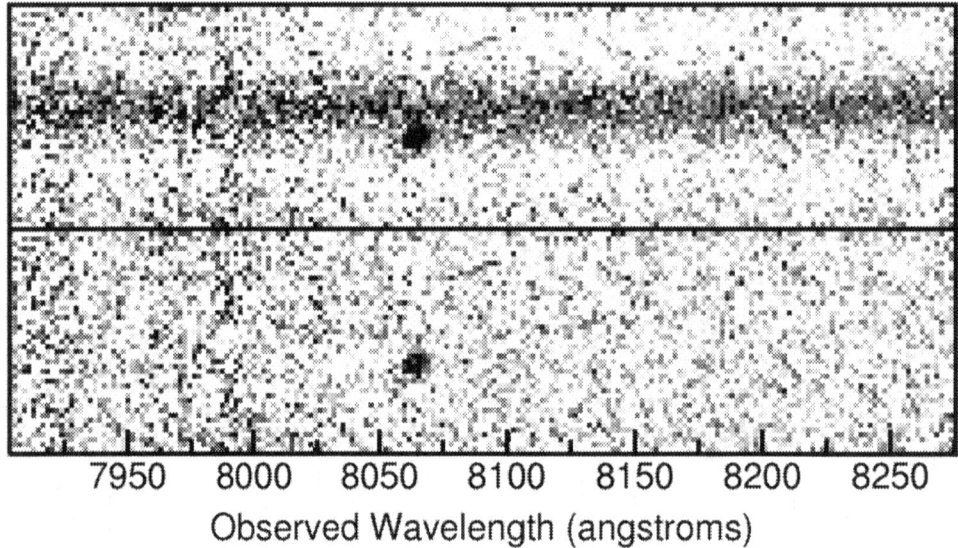

FIG. 2.8.— Discovery spectrogram for F 36246–1511, a solo emission line source interpreted as a Lyα–emitter at $z = 5.631$, lensed by an absorption–line galaxy (F 36247–1510) at $z = 0.641$. The top panel shows the raw spectrogram. The bottom panel shows the solo emission line after subtracting a Gaussian fit to the foreground continuum source. The fit was made to the continuum source blueward of the emission line so that after subtraction — assuming a locally flat spectrum for both sources — any remaining flux could be attributed to the solo line–emitter. The total exposure time is 5.4 ks. Each panel is 9″.5 in height. Since the discovery spectrum was obtained, we have targeted F 36246–1511 for an additional ∼ 25 ks of spectroscopy. The resulting composite spectrum, which confirms the high–redshift interpretation, will appear in a future work.

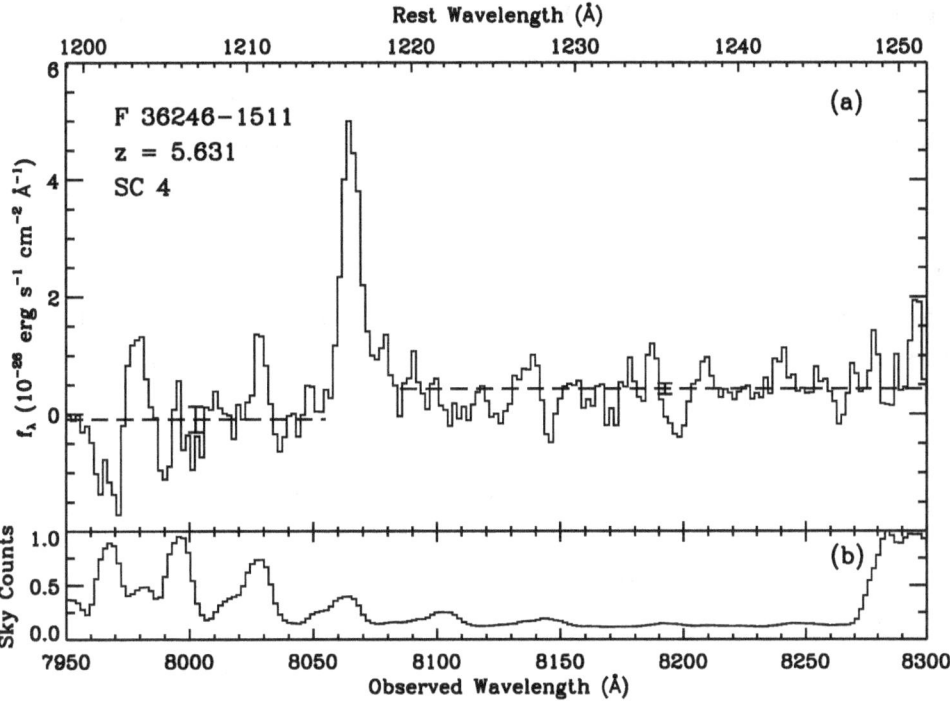

FIG. 2.9.— (a) The one–dimensional extracted spectrum of F 36246–1511. The dashed lines indicate the mean value of the spectrum over wavelengths lower than and higher than the emission line. The 1–sigma scatter in the two regions is 0.7×10^{-26} erg s^{-1} cm^{-2} Å^{-1} and 0.4×10^{-26} erg s^{-1} cm^{-2} Å^{-1}, respectively. The error bars indicate the 1–sigma scatter divided by the square root of the number of resolution elements in each region. Of course, the meaningfulness of these statistics is contingent on the source having a flat (or no) continuum on either side of the emission line. The total exposure time is 5.4 ks. The spectrum was smoothed using ~ 10 Å boxcar filter. (b) The normalized night–sky spectrum over the same observed wavelength range. For background–limited observations of faint objects, night–sky emission lines are the dominant source of noise.

likely high–redshift source, F 36246–1511 at $z = 5.631$, as illustrative of the situation.

F 36246–1511 was discovered in a 5400s exposure obtained on UT 19 February 1998. The source appeared as solo emission line spatially offset by $\sim 2''$ from an absorption–line galaxy (F 36247–1510; $z = 0.641$). A portion of the two–dimensional spectrogram, centered on the emission line, is shown in Figure 2.8. The top panel shows the original two–dimensional spectrogram; the continuum of the absorption–line galaxy and the spatially offset emission line can be readily seen. In the bottom panel, we have subtracted a Gaussian fit to the foreground continuum source. The fit was made to the continuum source only blueward of the emission line so that after subtraction — assuming a locally flat spectrum for both sources — any remaining flux could be attributed to the high–redshift candidate. In this fashion we hoped to isolate continuum flux from the high–redshift source and recover a continuum break, which would lend credence to the Lyα–interpretation. However, as can be seen in the one–dimensional extracted spectrum (Figure 2.9), the continuum break is of low significance relative to the noise.

As the emission line itself is not obviously asymmetric, the remaining evidence in favor of the Lyα–interpretation is two–fold. To begin, the observed frame equivalent width of the line is ~ 300 Å. This value exceeds the largest equivalent widths observed for other likely interpretations: 200 Å for the Hα+[N II] complex; 100 Å for [O III] $\lambda5007$; and 100 Å for [O II] (Stern & Spinrad 1999). Additionally, a faint source is visible in the Outer West I_{814} Flanking Field image (W96) located at the correct separation and orientation to be the progenitor of the solo emission line. Unfortunately, as the offset between the foreground continuum source and the high–redshift candidate is only $\sim 2''$, ground–based images are insufficient to resolve the two objects. Hence, the only available visual identification of the high–redshift candidate stems from the well–resolved but comparatively shallow single–orbit Flanking Field image.

Since the discovery spectrum was obtained, we have targeted F 36246–1511 for an additional ~ 25 ks of spectroscopy. The resulting composite spectrum confirms the $z = 5.631$ interpretation and will appear in a future work.

2.6   Conclusion

In the course of our on–going program to study distant galaxies in the HDF, we have produced as a fringe benefit a deep, serendipitous slit spectroscopy survey sensitive to a wide range of redshifts. Our catalog contains 74 serendipitously detected galaxies, 13 of which are galaxies in the central HDF which had no prior published spectroscopic redshift, 30 of which are galaxies located in the HDF Flanking Fields. Five of the serendipitously detected galaxies are members of a galaxy cluster at $z = 0.85$, and an additional five are candidate Lyα–emitters at $z \gtrsim 5$. The serendipitous sample demonstrates the redshift clustering behavior observed in other high–redshift samples. Moreover, our spectroscopic catalog indicates that photometric redshift techniques generally compare favorably with spectroscopic redshift determinations. As all of the spectra presented herein were obtained entirely without cost to the main observing campaign, the contribution made by this catalog to the rich database of observations of the HDF may be regarded as a testament to the persistent utility of serendipity in observational astronomy.

Acknowledgements

We are indebted to the expert staff of the Keck Observatory for their assistance in obtaining the data herein. It is a pleasure to thank T. Bida, W. Wack, J. Aycock R. Quick, T. Stickel, G. Punawai, R. Goodrich, R. Campbell, and B. Schaeffer for their invaluable assistance during Keck runs. We thank N. Brandt and B. Holden for many useful discussions concerning X–rays. We are grateful to M. Dickinson for acting as the steward of published redshifts in the HDF and for providing us with photometrically-selected targets; to J. Cohen for supporting LRIS and for making her own redshift survey available; to C. Steidel and A. Barger for releasing their optical and IR images of the HDF and its Flanking Fields; to A. Fernández–Soto for releasing and maintaining the interactive HDF catalog of photometric redshifts; and again to C. Steidel for supporting LRIS–B. SD is humbly indebted to JDS, GMR, MEE, JLW, and especially to EEBG, without whom this work would not have been possible. The work of DS was carried out at the Jet Propulsion Laboratory, California Institute of Technology, under contract with NASA. AJB was supported by a NICMOS postdoctoral fellowship while at Berkeley (NASA Grant NAG 5–3043), and a U.K. PPARC observational rolling grant postdoctoral position at the Institute of Astronomy in Cambridge (ref. no. PPA/G/O/1997/00793). HS gratefully acknowledges NSF grant AST 95–28536 for supporting much of the research presented herein. AD acknowledges partial support from NASA HF–01089.01–97A and from NOAO. NOAO is operated by AURA, Inc., under cooperative agreement with the NSF. This work made use of NASA's Astrophysics Data System Abstract Service.

TABLE 2.1

SPECTRAL CATEGORIES FOR SERENDIPITOUS DETECTIONS

Quality Class	Class Description
1	Multiple features
2	Solo line with continuum; assume [O II]
3	Solo line with continuum break; assume Lyα
4	Solo line with no continuum; assess imaging, if available
5	Continuum break; assess continuum strength blueward of break

TABLE 2.2

SERENDIPITOUSLY DETECTED GALAXIES IN THE CENTRAL HDF

ID[a]	I_{814}[a]	α_{J2000}[b]	δ_{J2000}[c]	z	SC[d]	References[e]	Comments[f]
1–95.0	24.07	$36^m45\rlap{.}''855$	$13'25\rlap{.}''81$	0.847	2	$\cdots$	[O II]
2–201.0	23.74	$36^m47\rlap{.}''178$	$13'41\rlap{.}''82$	1.313	1	$\cdots$	[O II], Mg II abs
2–173.0[g]	23.45	$36^m48\rlap{.}''474$	$13'16\rlap{.}''62$	0.474	1	$\cdots$	[O II], Ca II H,K abs
2–600.0[h]	25.59	$36^m49\rlap{.}''804$	$14'19\rlap{.}''15$	0.425	4	$\cdots$	[O II]
2–982.0	22.70	$36^m55\rlap{.}''528$	$13'53\rlap{.}''48$	1.144	1	C96, P97	[O II], Mg II abs
3–318.0	24.45	$36^m54\rlap{.}''805$	$12'58\rlap{.}''05$	0.851	2	$\cdots$	[O II]
3–342.0	24.57	$36^m58\rlap{.}''190$	$13'06\rlap{.}''58$	0.475	1	$\cdots$	[O III]a,b, Hβ
3–430.1	24.30	$36^m56\rlap{.}''603$	$12'52\rlap{.}''70$	1.233	2	C00	[O II]
3–773.0	22.46	$36^m57\rlap{.}''214$	$12'25\rlap{.}''83$	0.563	1	C96	[O II], [O III]a,b
3–863.0	23.39	$36^m58\rlap{.}''649$	$12'21\rlap{.}''72$	0.682	1	C96	[O II], [O III]a,b
4–131.0	24.91	$36^m49\rlap{.}''365$	$12'14\rlap{.}''64$	0.934	2	$\cdots$	[O II]
4–236.0[i]	28.26	$36^m47\rlap{.}''838$	$12'18\rlap{.}''30$	0.102	4	$\cdots$	[O III]b
4–402.1	24.96	$36^m43\rlap{.}''822$	$12'51\rlap{.}''96$	1.013	2	$\cdots$	[O II]
4–402.3	21.13	$36^m43\rlap{.}''964$	$12'50\rlap{.}''13$	0.557	2	C96, C00	[O II]
4–416.0	24.38	$36^m46\rlap{.}''555$	$12'03\rlap{.}''09$	0.454	1	C00	[O II], Hβ
4–430.0	23.30	$36^m44\rlap{.}''181$	$12'40\rlap{.}''39$	0.873	4	C96, C00	[O II]
4–471.0	21.93	$36^m46\rlap{.}''511$	$11'51\rlap{.}''32$	0.503	4	C96	[O II]
4–491.0[j]	24.86	$36^m43\rlap{.}''253$	$12'38\rlap{.}''86$	2.442	3	$\cdots$	Lyα
4–493.0	21.74	$36^m43\rlap{.}''156$	$12'42\rlap{.}''20$	0.849	1	C96, C00	Ca II H,K abs, D4000, G band
4–565.0	22.68	$36^m43\rlap{.}''627$	$12'18\rlap{.}''25$	0.749	1	C96, C00	[O II], [O III]b
4–639.1	24.65	$36^m41\rlap{.}''712$	$12'38\rlap{.}''75$	2.592	3	S96, C00	Lyα
4–658.0	24.77	$36^m44\rlap{.}''734$	$11'43\rlap{.}''77$	0.558	1	$\cdots$	[O II], [O III]a,b

Continued on next page...

TABLE 2.2—*Continued*

ID[a]	I_{814}[a]	α_{J2000}[b]	δ_{J2000}[c]	z	SC[d]	References[e]	Comments[f]
4–727.0	23.00	$36^m43\rlap{.}''409$	$11'51\rlap{.}''57$	1.238	2	C00	[O II]
4–937.0	25.09	$36^m42\rlap{.}''284$	$11'26\rlap{.}''18$	0.559	1	$\cdots$	[O II], [Ne III]
4–948[k]	24.99	$36^m41\rlap{.}''427$	$11'42\rlap{.}''89$	1.524	2	$\cdots$	[O II]

[a]Object IDs and I_{814} magnitudes are from Williams et al. (1996).

[b]Add 12 hours to the right ascension.

[c]Add 62 degrees to the declination.

[d]See §2.3.2 and Table 2.1.

[e] *References* lists spectroscopic redshifts already in the literature. The following abbreviations are used: C96 = Cohen et al. (1996), C00 = Cohen et al. (2000), P97 = Phillips et al. (1997), S96 = Steidel et al. (1996a).

[f]The oxygen emission lines are abbreviated: [O II] = [O II]; [O III]a = [O III] $\lambda4959$; [O III]b = [O III] $\lambda5007$.

[g]Listed without a redshift as H36485_1317 in Cohen et al. (2000). Redshift identification tentative; weak detection.

[h]Redshift identification tentative. Weak detection consistent with [O II]–interpretation of solo line; possible detection of very faint additional lines is roughly consistent with [O III] $\lambda5007$–interpretation.

[i]Redshift identification tentative. Weak detection. Object colors (see W96) are not consistent with Lyα–interpretation of solo line; [O II]–interpretation suggests presence of [O III] $\lambda5007$ at $\lambda_{\text{obs}} = 7415$ Å, which is not detected; [O III] $\lambda5007$–interpretation suggests presence of Hα at $\lambda_{\text{obs}} = 7232$ Å, which may be very weakly detected.

[j]Listed as NICMOS #850 with $z_{\text{phot}} = 2.40$ (Dickinson 2001, private communication).

[k]The data given are for 4–948.1111, a daughter object likely to be a part of the system formed by 4–948.2, 4–948.11, 4–948.111, 4–948.112, 4–948.1112, 4–948.11111, and 4–948.11112. This system is distinct from that formed by 4–948.0, 4–948.1, and 4–948.12, which has a redshift of $z = 0.585$ (Phillips et al. 1997; Cohen et al. 2000).

TABLE 2.3

SERENDIPITOUSLY DETECTED GALAXIES OUTSIDE OF THE CENTRAL HDF

ID	I_{AB}[a]	α_{J2000}[b]	δ_{J2000}[c]	z	SC[d]	FF[e]	Comments[f]
F 36179–1635	20.1	$36^m 17\rlap{.}''97$	$16'35\rlap{.}''0$	0.681	1	$\cdots$	[O II], [O III]a,b, Ca II H,K abs
F 36184–1601	22.3	$36^m 18\rlap{.}''43$	$16'01\rlap{.}''6$	0.797	1	$\cdots$	[O II], [O III]a,b, Hβ
F 36191–6217	> 25.0	$36^m 19\rlap{.}''12$	$17'04\rlap{.}''2$	3.896	4	$\cdots$	Lyα; pstn. from spectrum
F 36197–1601	22.9	$36^m 19\rlap{.}''78$	$16'01\rlap{.}''3$	1.345	1	$\cdots$	[O II], Mg II abs
F 36218–1513	> 25.0	$36^m 21\rlap{.}''87$	$15'13\rlap{.}''7$	5.767	4	OW	Lyα; pstn. from spectrum
F 36219–1516	24.4	$36^m 21\rlap{.}''91$	$15'16\rlap{.}''8$	4.890	3	OW	Lyα
F 36220–1459	22.9	$36^m 22\rlap{.}''04$	$14'59\rlap{.}''7$	0.849	2	OW	[O II]
F 36240–1516	23.3	$36^m 24\rlap{.}''05$	$15'16\rlap{.}''2$	0.796	2	OW	[O II]
F 36241–1514	22.7	$36^m 24\rlap{.}''18$	$15'14\rlap{.}''5$	0.222	1	OW	Hα, [O III]b, Hβ
F 36246–1511	> 25.0	$36^m 24\rlap{.}''61$	$15'11\rlap{.}''9$	5.631	4	OW	Lyα; pstn. from spectrum
F 36247–1510	20.1	$36^m 24\rlap{.}''70$	$15'10\rlap{.}''5$	0.641	1	OW	Ca II H,K, Hδ abs, D4000
F 36249–1834[g]	$\cdots$	$36^m 24\rlap{.}''92$	$18'34\rlap{.}''1$	0.852	2	$\cdots$	[O II]
F 36255–1510	22.7	$36^m 25\rlap{.}''50$	$15'10\rlap{.}''7$	0.680	2	OW	[O II]
F 36265–1443	24.2	$36^m 26\rlap{.}''58$	$14'43\rlap{.}''9$	0.625	1	OW	[O II], [O III]a,b, Hβ, Hγ
F 36270–1509	20.7	$36^m 27\rlap{.}''04$	$15'09\rlap{.}''4$	0.794	1	OW	Ca II H,K abs
F 36279–1507	21.4	$36^m 27\rlap{.}''98$	$15'07\rlap{.}''8$	0.680	2	OW	[O II]
F 36279–1750[g]	$\cdots$	$36^m 27\rlap{.}''97$	$17'50\rlap{.}''4$	4.938	4	$\cdots$	Lyα; pstn. from spectrum
F 36289–1752[g]	$\cdots$	$36^m 28\rlap{.}''93$	$17'52\rlap{.}''7$	1.592	2	$\cdots$	[O II]
F 36316–1604	21.1	$36^m 31\rlap{.}''65$	$16'04\rlap{.}''1$	0.785	2	$\cdots$	[O II]
F 36339–1604	22.4	$36^m 33\rlap{.}''97$	$16'04\rlap{.}''7$	0.834	1	$\cdots$	[O II], [O III]a,b
F 36348–1628	22.1	$36^m 34\rlap{.}''87$	$16'28\rlap{.}''4$	0.847	1	$\cdots$	[O II], Ca II H,K abs
F 36356–1424[h]	23.1	$36^m 35\rlap{.}''59$	$14'24\rlap{.}''0$	2.011	1	IW	See §2.5.4
F 36361–1656	20.9	$36^m 36\rlap{.}''16$	$16'56\rlap{.}''9$	0.488	1	$\cdots$	[O II], Hα
F 36362–1709	21.8	$36^m 36\rlap{.}''22$	$17'09\rlap{.}''3$	0.945	2	$\cdots$	[O II]
F 36367–1604	22.6	$36^m 36\rlap{.}''77$	$16'04\rlap{.}''8$	0.851	5	$\cdots$	D4000
F 36376–1047	22.3	$36^m 37\rlap{.}''64$	$11'47\rlap{.}''8$	0.880	2	SW	[O II]
F 36376–1453	22.4	$36^m 37\rlap{.}''63$	$14'53\rlap{.}''7$	4.886	4	IW	Lyα; visual ID uncertain
F 36382–1053	23.7	$36^m 38\rlap{.}''20$	$10'53\rlap{.}''0$	0.766	2	SW	[O II]
F 36382–1605	21.2	$36^m 38\rlap{.}''22$	$16'05\rlap{.}''1$	0.852	1	$\cdots$	[O II], D4000
F 36387–1059	24.8	$36^m 38\rlap{.}''75$	$11'59\rlap{.}''3$	3.956	4	SW	Lyα
F 36397–1547	21.0	$36^m 39\rlap{.}''76$	$15'47\rlap{.}''9$	0.847	1	$\cdots$	Ca II H,K abs, D4000
F 36398–1601	22.8	$36^m 39\rlap{.}''83$	$16'01\rlap{.}''6$	0.843	5	$\cdots$	D4000
F 36405–1334	24.1	$36^m 40\rlap{.}''51$	$13'34\rlap{.}''9$	3.826	4	IW	Lyα
F 36417–1437	23.4	$36^m 41\rlap{.}''72$	$14'37\rlap{.}''7$	0.940	2	IW	[O II]
F 36452–1108	23.3	$36^m 45\rlap{.}''24$	$11'08\rlap{.}''8$	0.512	1	SE	[O II], Hβ, [O III]a,b
F 36466–1517	24.9	$36^m 46\rlap{.}''68$	$15'17\rlap{.}''2$	0.652	2	NW	[O II]; visual ID uncertain
F 36488–1500	> 25.0	$36^m 48\rlap{.}''87$	$15'00\rlap{.}''6$	2.924	4	NW	Lyα; pstn. from spectrum
F 36488–1502[i]	24.4	$36^m 48\rlap{.}''87$	$15'02\rlap{.}''5$	3.111	3	NW	Lyα
F 36490–1512	22.7	$36^m 49\rlap{.}''07$	$15'12\rlap{.}''4$	0.458	1	NW	[O II], Hβ

Continued on next page...

TABLE 2.3—*Continued*

ID	I_{AB}[a]	α_{J2000}[b]	δ_{J2000}[c]	z	SC[d]	FF[e]	Comments[f]
F 36490–1620	21.8	$36^m49\rlap{.}{''}08$	$16'20\rlap{.}{''}8$	0.501	2	NW	[O II]
F 36492–1645	23.4	$36^m49\rlap{.}{''}25$	$16'45\rlap{.}{''}7$	0.536	4	NW	[O II]
F 36568–1353	25.0	$36^m56\rlap{.}{''}88$	$13'53\rlap{.}{''}6$	3.43:	5	NE	Ly break
F 37043–1335	22.9	$37^m04\rlap{.}{''}35$	$13'35\rlap{.}{''}3$	0.592	1	IE	Ca II H,K abs; visual ID uncertain
F 37051–1210	22.5	$37^m05\rlap{.}{''}18$	$12'10\rlap{.}{''}7$	0.387	1	IE	Hα, [O III]a,b, Hβ
F 37069–1208	23.7	$37^m06\rlap{.}{''}98$	$12'08\rlap{.}{''}1$	0.693	1	IE	[O II], Hβ, [O III]a,b
F 37098–1400	24.8	$37^m09\rlap{.}{''}80$	$14'00\rlap{.}{''}2$	3.910	3	$\cdots$	Lyα
F 37131–1333	21.9	$37^m13\rlap{.}{''}11$	$13'33\rlap{.}{''}8$	0.842	1	$\cdots$	[O II], [O III]a,b
F 37138–1335	21.5	$37^m13\rlap{.}{''}88$	$13'35\rlap{.}{''}2$	0.776	2	$\cdots$	[O II]
F 37180–1248	22.4	$37^m18\rlap{.}{''}06$	$12'48\rlap{.}{''}2$	0.908	2	OE	[O II]

[a]Isophotal magnitude.

[b]Add 12 hours to the right ascension.

[c]Add 62 degrees to the declination.

[d]See §2.3.2 and Table 2.1.

[e]Indicated galaxy is located in one of the HDF Flanking Field observations (see Williams et al. 1996, Table 2): OW = Outer West; SW = South West; IW = Inner West; SE = South East; NW = North West; NE = North East; IE = Inner East; OW = Outer East.

[f]Oxygen emission lines are abbreviated: [O II] = [O II]; [O III]a = [O III] $\lambda4959$; [O III]b = [O III] $\lambda5007$.

[g]Indicated galaxy falls outside of the Hawaii 2.2m I–band image of Barger et al. (1999); the identification was made in our own 70 minute R–band image obtained with ESI (where possible); I_{AB} magnitudes are not available.

[h]Optical ID for X–ray source CXOHDFN J123635.6+621424 (Hornschemeier et al. 2001). See §2.5.4.

[i]Redshift identification tentative. See discussion of SC 3 in §2.3.2.

TABLE 2.4

SUMMARY OF PROPERTIES OF SPECTROSCOPIC MEMBERS OF CLG 1236+6215

ID[a]	α_{J2000}[b]	δ_{J2000}[c]	z	$(V-I)_{\text{AB}}$	Radius[d]
F 36348–1628	$36^m34\!''\!.87$	$16'28\!''\!.4$	0.847	1.9	$9\!''\!.8$
F 36367–1604	$36^m36\!''\!.77$	$16'04\!''\!.8$	0.851	2.4	$42\!''\!.7$
F 36382–1605	$36^m38\!''\!.22$	$16'05\!''\!.1$	0.852	2.9	$10\!''\!.2$
F 36397–1547	$36^m39\!''\!.76$	$15'47\!''\!.9$	0.847	2.6	$10\!''\!.5$
F 36398–1601	$36^m39\!''\!.83$	$16'01\!''\!.6$	0.843	2.5	$16\!''\!.8$
C 36392–1623	$36^m39\!''\!.22$	$16'23\!''\!.4$	0.850	2.4	$28\!''\!.2$
C 36421–1545	$36^m42\!''\!.16$	$15'45\!''\!.2$	0.857	1.8	$25\!''\!.8$
C 36435–1532	$36^m43\!''\!.50$	$15'32\!''\!.2$	0.847	2.7	$40\!''\!.3$

[a]Entries beginning with F are galaxies described in this catalogue. Entries beginning with C are described in Cohen et al. (2000).

[b]Add 12 hours to the right ascension.

[c]Add 62 degrees to the declination.

[d]*Radius* indicates the angular distance of the galaxy from the nominal cluster center: $\alpha = 12^h36^m39\!''\!.6$ $\delta = +62°15'54''$ (J2000).

Chapter 3

A Galactic Wind at z = 5.190

A version of this chapter was previously published in *The Astrophysical Journal* (Dawson, S., Spinrad, H., Stern, D., Dey, A., van Breugel, W., de Vries, W., & Reuland, M. 2002, ApJ, 570, 92). Reproduced by permission of the AAS.

Abstract

We report the serendipitous detection in high–resolution optical spectroscopy of a strong, asymmetric Lyα emission line at $z = 5.190$. The detection was made in a 2.25 hour exposure with the Echelle Spectrograph and Imager on the Keck II telescope through a spectroscopic slit of dimensions $1'' \times 20''$. The progenitor of the emission line, J123649.2+621539 (hereafter ES1), lies in the Hubble Deep Field North West Flanking Field where it appears faint and compact, subtending just $0\rlap{.}''3$ (FWHM) with $I_{AB} = 25.4$. The ES1 Lyα line flux of 3.0×10^{-17} ergs cm^{-2} s^{-1} corresponds to a luminosity of 9.0×10^{42} ergs s^{-1}, and the line profile shows the sharp blue cut–off and broad red wing commonly observed in star–forming systems and expected for radiative transfer in an expanding envelope. We find that the Lyα profile is consistent with a galaxy–scale outflow with a velocity of $v > 300$ km s^{-1}. This value is consistent with wind speeds observed in powerful local starbursts (typically 10^2 to 10^3 km s^{-1}), and compares favorably to simulations of the late–stage evolution of Lyα emission in star–forming systems. We discuss the implications of this high–redshift galactic wind for the early history of the evolution of galaxies and the intergalactic medium, and for the origin of the UV background at $z > 3$.

3.1 Introduction

Following the epoch of recombination, the Universe settled into the comparatively dormant dark ages, during which the primordial glow had begun to fade but the first present–day astronomical objects had yet to form. This tranquil epoch proved short–lived, however, as the formation of the first stars and quasars ushered in the first era of cosmological heating and enrichment at $z < 20$ (e.g. Gnedin & Ostriker 1997; Haiman & Loeb 1997, 1998; Ostriker & Gnedin 1996; Valageas & Silk 1999). Evidence of these processes in the form of galaxy–scale outflows is abundant in spectroscopy of the high–redshift Universe. Both optical/IR spectra of the $z \sim 3$ Lyman–break population (e.g. Pettini et al. 2001) and optical spectra of lensed Lyα–emitting galaxies at $z > 4$ (Frye et al. 2002) show metal absorption lines which are blueshifted by hundreds of km s^{-1} with respect to the stellar rest frame of the galaxy, and Lyα emission lines which are shifted similarly redward. These observations, as well as the characteristic P–Cygni profile of the Lyα emission lines (e.g. Bunker, Moustakas, & Davis 2000; Dey et al. 1997, 1998; Dickinson 1998; Ellis et al. 2001, Lowenthal et al. 1997; Weymann et al. 1998) paint a coherent picture of optically thick expanding regions surrounding star–forming galaxies, most naturally driven by the starbursts that render them visible in the first place (e.g. Heckman et al. 2000, and references therein).

We present a direct observation of such an outflow at $z = 5.190$ in high–resolution optical spectroscopy of the serendipitously detected star–forming galaxy J123649.2+621539 (hereafter ES1, for Echelle Spectrograph and Imager serendipitous detection number one). The sharp blue cut–off and broad red wing of the ES1 Lyα emission line are consistent with the profile expected for the transfer of line radiation in an expanding envelope (e.g. Surdej 1979). The suggested outflow velocity of $v > 300$ km s^{-1} is in broad agreement with simulations of the late evolution of Lyα emission and absorption in star–forming galaxies (e.g. Tenorio–Tagle et al. 1999), and is consistent with observations of powerful nearby starbursts (e.g. Heckman, Armus, & Miley 1990). The spectrum of ES1 therefore presents evidence for both a high star–formation rate and a high–redshift, starburst–driven galactic wind, fitting well within the expectations of current models for the early history of galaxy formation.

In §3.2 we discuss the detection of ES1 and we give a description of the spectrum and the available archival imaging. In §3.3 we detail the properties of the ES1 Lyα emission line, and we present a model for the emission line profile consistent with the expanding shell scenario introduced above. We conclude in §3.4 with a discussion of the implications of the evidence for a strong outflow in ES1 for both the evolution of galaxies and the intergalactic medium (IGM) at high redshift, and for the origin of the UV background at $z > 3$. Throughout this paper we adopt the currently favored Λ–cosmology of $\Omega_{\mathrm{M}} = 0.35$ and $\Omega_{\Lambda} = 0.65$, with $H_0 = 65$ km s^{-1} Mpc^{-1} (e.g. Riess et al. 2001). At $z = 5.190$, such a universe is only 1.10 Gyr old — corresponding to a look–back time of 92.1% of the age of the Universe — and an angular size of $1''\!.0$ corresponds to 6.31 kpc.

3.2   Observation and Data Reduction

ES1 was detected in a 2.25 hour exposure made with the Echelle Spectrograph and Imager (ESI; Sheinis et al. 2000) at the Cassegrain focus of the Keck II telescope on UT 2001 February 25. The instrument was configured in its medium–resolution echellete mode with a spectroscopic slit of dimensions $1'' \times 20''$, yielding a spectral resolution of ~ 2 Å (78 km s^{-1}) at 7500 Å. The 2.25 hour exposure was broken into four integrations of 1800 seconds and one integration of 900 seconds; we performed $3''$ spatial offsets between each integration to facilitate the removal of fringing at long wavelengths. We used IRAF[1] (Tody 1993) to process the echellogram, following standard slit spectroscopy procedures (e.g. Massey, Valdes, & Barnes 1993)[2]. Some aspects of treating the ten individual orders of the spectrum were facilitated by the software package BOGUS[3], created by D. Stern, A.J. Bunker, and S.A. Stanford. Wavelength calibrations were performed in the standard fashion using Xe, HgNe, and CuAr arc lamps; we employed telluric sky lines to adjust the wavelength zero–point. The night was near photometric with $0''\!.6$ seeing, and we performed flux calibrations with observations of standard stars from Massey & Gronwall (1990) taken with the instrument in the same configuration as the target observation. A portion of one order of the discovery spectrum, centered on the ES1 emission line, is shown in Figure 3.1.

The target object for this observation was a Lyman–break galaxy at $z = 3.125$ (D16; Steidel 2001, private communication), located in the Hubble Deep Field North West Flank-

---

[1]IRAF is distributed by the National Optical Astronomy Observatories, which are operated by the Association of Universities for Research in Astronomy, Inc., under cooperative agreement with the National Science Foundation.

[2]*A   User's   Guide   to   Reducing   Slit   Spectra   with   IRAF*,   available   online   at http://iraf.noao.edu/iraf/web/docs/spectra.html

[3]BOGUS is available online at http://zwolfkinder.jpl.nasa.gov/~stern/homepage/bogus.html.

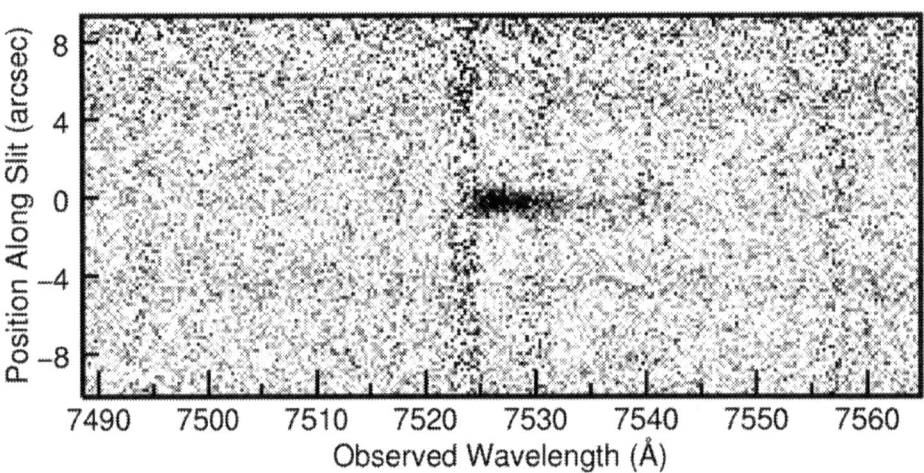

FIG. 3.1.— A portion of one order of the ES1 discovery spectrum. The dispersion axis is horizontal; the spatial axis is vertical. The vertical features flanking the emission line are remnants of the OH and O_2 night–sky emission lines at 7524 Å and 7531 Å, respectively. The continuum of the target galaxy, D16, is barely visible near the slit position $-4''$. See §3.2 for a description of the observation.

ing Field (Williams et al. 1996). ES1 was fortuitously placed on the spectroscopic slit roughly $4''$ south of D16, along the parallactic angle of $150°$ (Figure 3.2). Upon making the observation, we immediately noticed a strong, serendipitously detected emission line near 7527 Å whose asymmetric profile suggested redshifted Lyα at $z = 5.190$. As the spectrograph configuration for the discovery spectrum covered only 20 arcsec^2 — suggesting a surface density of high redshift Lyα–emitters roughly 30 to 90 times in excess of current estimates (e.g. Cowie & Hu 1998; Dawson et al. 2001; Stern & Spinrad 1999; Thompson, Weymann, & Storrie–Lombardi 2001) — we consider the detection of ES1 to be highly providential.

Careful inspection of the reduced spectrum failed to reveal emission lines at other wavelengths. Though the galaxy continuum is faintly detected redward of the emission line, the observation did not achieve sufficient signal–to–noise to confirm the presence or absence of interstellar absorption lines. The non–detection of high–ionization state emission lines typically observed in AGN spectra, e.g. N 5 λ1240, C 4 λ1549, or He 2 λ1640, strongly suggests that the Lyα emission in ES1 is due to the Lyman continuum

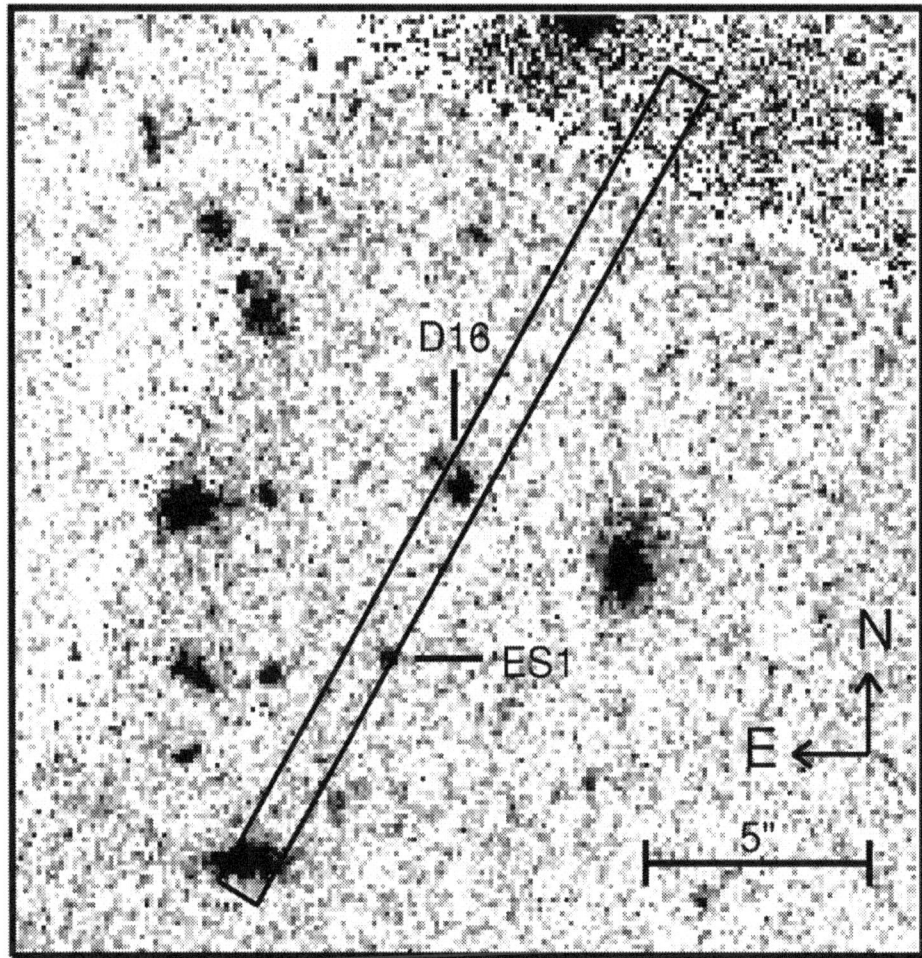

FIG. 3.2.— Central region of the *HST* I_{814} mosaic of the Hubble Deep Field North West Flanking Field with a projection of the spectroscopic slit. The target galaxy (D16) at $\alpha = 12^h 36^m 49\rlap{.}{''}0$, $\delta = +62°15'43''$ (J2000) and the serendipitously detected galaxy (ES1) at $\alpha = 12^h 36^m 49\rlap{.}{''}2$, $\delta = +62°15'39''$ (J2000) are indicated. The panel measures 20'' square and the slit dimensions are 1'' × 20''. The mosaic and the astrometry therein were provided by Dickinson (2001, private communication). The slit position was determined by offsetting from the set–up star used in the original observation (not shown), thereby nullifying errors in absolute astrometry.

flux of OB stars, rather than the hard UV spectrum of an AGN. For $z = 5.190$, the nebular emission lines typically associated with star–forming galaxies are inaccessible to optical spectroscopy.

We were fortunate that ES1 is located in the HDF North West Flanking Field. Figure 3.2 displays a portion of the single–orbit *Hubble Space Telescope* (hereafter *HST*) I_{814} image of that region with a projection of the spectroscopic slit. ES1 is faint and compact, typical for a very distant galaxy (Steidel et al. 1996b). By running the source extraction algorithm SExtractor (Bertin & Arnouts 1996) on the Flanking Field image and employing the conversion from data number to AB magnitude given in Williams et al. (1996), we determine an isophotal magnitude for ES1 of $I_{AB} = 25.4 \pm 0.2$. This is challengingly faint, comparable to the first detected galaxies at $z > 5$, e.g. HDF 3–951 with $I_{814} = 25.6$ (Spinrad et al. 1998). Moreover, as an appreciable fraction of the I–band radiation for the source is from the Lyα emission line, the true continuum magnitude must be still fainter than that determined from the source extraction. ES1 appears marginally resolved, subtending $0\rlap{.}''3$ (FWHM) on the Flanking Field image, while stars on the same image have a FWHM near $0\rlap{.}''2$. Thus, the physical size of the emitting region appears to be only ~ 4 kpc in diameter.

We present ground–based V, R, I, and z images of ES1 in Figure 3.3 and we give the ground–based photometry in Table 3.2. The V and I images are from the Canada France Hawaii Telescope imaging campaign of Barger et al. (1999); the R and z images are from our own Keck imaging campaign of the HDF and its environs (see Stern et al. 2000b). The fact that ES1 is not detectable in the V or R bands is characteristic of high–redshift galaxies, where intervening neutral hydrogen (the Lyman forests) severely attenuates the continuum signal blueward of Lyα (Madau 1995; Steidel et al. 1996b; Stern & Spinrad 1999). The fact that ES1 is not detectable in the z band is due only to the comparative shallowness of the z band image. Based on the model spectrum described in section § 3.3.2 and assuming that ES1 has a flat spectrum in f_ν at wavelengths longer than the emission line, we expect a Vega–based continuum magnitude in the z band of $z \gtrsim 27.0$. This value is almost five times as faint as the 3σ limiting magnitude in that image (see Table 3.2).

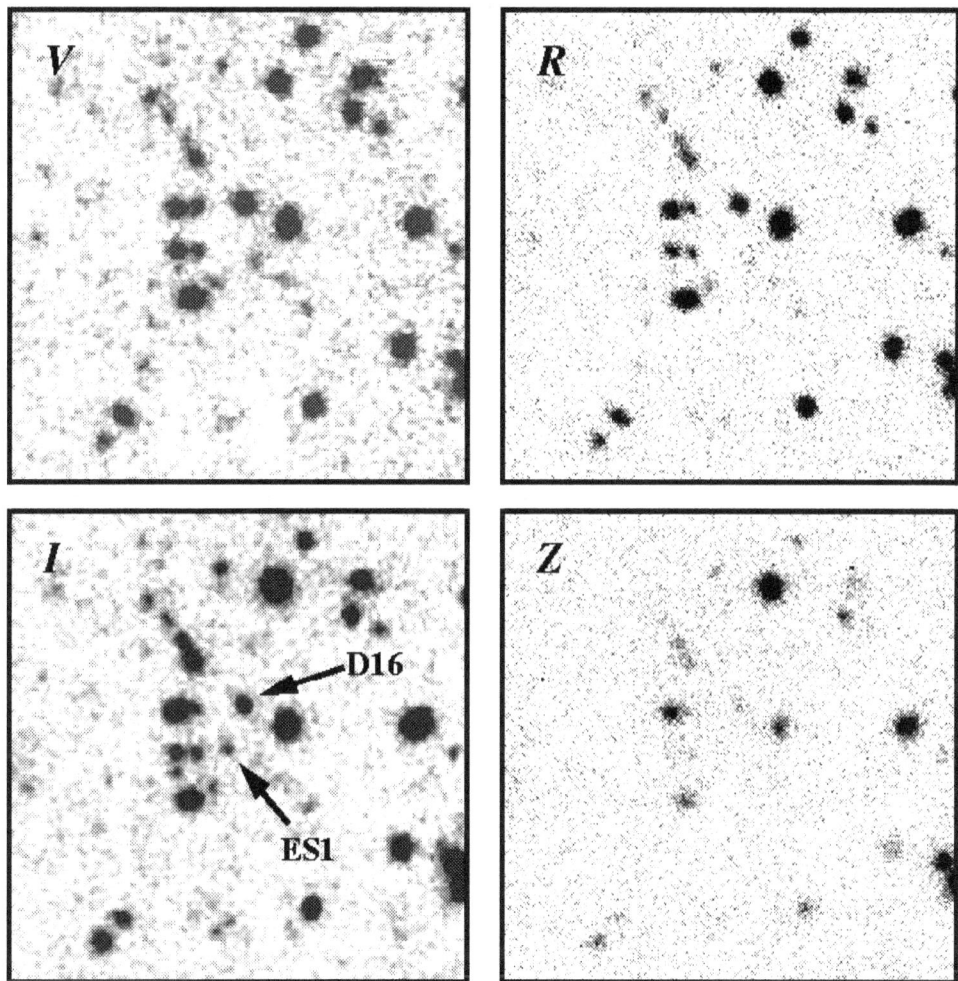

FIG. 3.3.— Ground–based supporting imaging for ES1. The V and I images are from Barger et al. (1999); the R and z images are from our own imaging campaign of the HDF and its environs (see Stern et al. 2000b). The fields are $1'$ square; North is up and East is to the left. Notice that ES1 is not detectable in the V or R bands but is seen in the I band. This is characteristic of high–redshift galaxies, where intervening neutral hydrogen (the Lyman forests) severely attenuates the continuum signal blueward of Lyα. The fact that ES1 is not detectable in the z band is due only to the shallowness of that image. We expect the z band continuum magnitude for ES1 to be $z > 27.0$; the 3σ limiting magnitude of the z band image is $z = 25.2$. See Table 3.2 for a summary of the ground–based photometry.

3.3   Properties of the Lyα Emission Line

3.3.1   The Emission Line Luminosity & Equivalent Width

In the fashionable cosmology $\Omega_M = 0.35$, $\Omega_\Lambda = 0.65$, the ES1 Lyα flux of 3.0×10^{-17} ergs cm^{-2} s^{-1} corresponds to a line luminosity of 9.0×10^{42} ergs s^{-1}. For the prescription given in Dey et al. (1998) assuming no dust aborption and negligible extinction, this luminosity corresponds to a star formation rate (SFR) of $\sim$ 10 M$_\odot$ yr^{-1}. As is evident in Figure 3.4, these values are typical of Lyα–emitting galaxies at high redshift ($z > 3$) discovered serendipitously (e.g. Dawson et al. 2001; Dey et al. 1998; Manning et al. 2000; Spinrad et al. 1999; Stern & Spinrad 1999) or in narrow band surveys (e.g. Hu et al. 1999; Rhoads et al. 2000). Of course, such galaxies may represent rare systems at the high–luminosity tip of an unexplored underlying population, and the observed Lyα luminosities may accordingly be governed by a selection effect. This point is borne out by the faint system discovered in the blind spectroscopic survey of well–constrained lensing clusters (Ellis et al. 2001), which is at a higher redshift ($z = 5.576$) and is an order of magnitude less luminous in Lyα than ES1.

We estimate a rest–frame equivalent width for the emission line of $W_\lambda^{\rm rest} = 120 \pm 40$ Å based on the emission line model described in the following section. This result should be treated with a degree of circumspection, however. The foremost caveat is that both the total line flux and the continuum level which enter the calculation were obtained from the model spectrum before attenuation by neutral hydrogen absorption. In this manner we attempted to circumvent the ambiguity in deriving an equivalent width from a strongly P–Cygni line profile coupled with a pronounced continuum break, essentially arriving at a theoretical equivalent width from the spectrum as it would appear if not truncated by foreground absorption. As a second caveat, the continuum redward of the emission line is not well–detected, with a significance of far less than 1σ. Together, we expect these two effects to cause an over-estimation of $W_\lambda^{\rm rest}$. However, this tendency toward over-estimation may be offset by the fact that this is a serendipitous detection, so ES1 was not well centered in the spectroscopic slit (see Figure 3.2). As such, we have been conservative in our estimate of the uncertainty in $W_\lambda^{\rm rest}$, which includes sky noise, the uncertainty in the continuum level redward of the emission line in the extracted spectrum, and the fit errors in the model spectrum. Even so, with $W_\lambda^{\rm rest} = 120 \pm 40$ Å, ES1 figures in the top 2% of Lyman–break galaxies with Lyα in emission in the near–complete continuum–selected

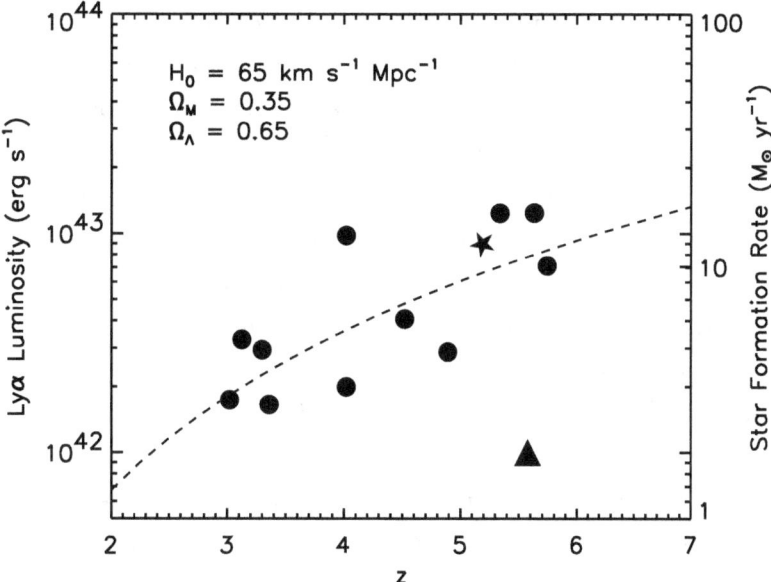

FIG. 3.4.— Lyα emission line luminosity vs. redshift for 13 Lyα–emitting galaxies in an $\Omega_M = 0.35$, $\Omega_\Lambda = 0.65$ cosmology. ES1 is indicated with a star. The SFR scale has been adopted from Dey et al. (1998). The sample of galaxies represented by circles was compiled from Dawson et al. (2001), Dey et al. (1998), Hu, McMahon, & Cowie (1999), Manning et al. (2000), Rhoads et al. (2000), Spinrad et al. (1999), Stanford (2001, private communication), and Stern & Spinrad (1999). The galaxy represented by the triangle is from Ellis et al. (2001). The dashed line indicates the limiting sensitivity to line flux in the Large Area Lyman Alpha Survey ($\sim 2 \times 10^{-17}$ erg cm^{-2} s^{-1}; Rhoads et al. 2000).

sample of Steidel et al. (2000).

3.3.2   Modeling the Emission Line

The asymmetric Lyα emission lines commonly observed in high–redshift starburst galaxies are generally ascribed to the interaction of Lyman continuum photons generated by newborn OB associations with a galaxy–scale expanding shell of neutral hydrogen. For a sufficiently massive starburst, the hot ionized gas created in the vicinity of the stars vents into the halo of the galaxy, where it sweeps up neutral hydrogen into an optically thick shell. Recombination in the ionized gas converts Lyman continuum photons escaping from the surface of the hot stars into line photons. Then, from the vantage of an observer, the near side of the expanding shell absorbs photons on the blue side of the resonant Lyα emission line, causing a flux decrement on what would otherwise be the blue wing of the Lyα emission. The far side of the shell back–scatters Lyα photons into the observer's line–of–sight; as these photons are offset redward by hundreds of km s^{-1} from both the rest frame of the galaxy and the approaching side of the neutral shell, they escape the galaxy and impose a pronounced red wing on the emission line profile. The net effect is to create the P–Cygni profile ubiquitous in observations of expanding shells.

The comparatively high signal–to–noise of the ES1 Lyα detection created the opportunity for probing this emission line structure with a simple model. In accordance with the scenario described above, we fit the Lyα emission feature with three components: (1) a comparatively large amplitude, narrow Gaussian intended to model line radiation generated by recombination in the ionized hydrogen; (2) a small amplitude, broad Gaussian intended to model the red wing of line photons back–scattered off the far side of the expanding shell; and (3) a Voigt absorption profile intended to model the blue decrement caused by the absorption of line photons by the near side of the shell. We modeled the weak continuum with a constant ($f_\lambda \propto \lambda^0$) baseline.

To account for the continuum decrement caused by line blanketing in the Lyα forest, we attenuated the model spectrum by a transmission profile adopted from Madau (1995), where the optical depth due to the combined effect of many Lyα absorption lines is given as

$$\tau = 0.0036 \left(\frac{\lambda}{\lambda_\alpha}\right)^{3.46}, \tag{3.1}$$

with $\lambda_\alpha = 1216$ Å. At high redshifts ($z > 4.5$), absorption by metal systems makes a non–negligible contribution to cosmic opacity. Hence, we accounted for the combined effect of metal lines with the additional factor also given in Madau (1995),

$$\tau = 0.0017 \left(\frac{\lambda}{\lambda_\alpha}\right)^{1.68}. \tag{3.2}$$

Neither Lyman series line blanketing nor continuum absorption from neutral hydrogen were included, as both these effects fall shortward of the wavelength range of interest.

Figure 3.5 shows the minimum–χ^2 model resulting when the centroids, amplitudes, and widths of each of the model components are left unconstrained. We weighted the χ^2–fit by the error spectrum shown in the figure; this had the effect of diminishing the contribution to the fit by pixels which fall on OH and O_2 night sky emission lines (which are the dominant source of error in low signal–to–noise spectra over the wavelengths of interest). We note that the residuals are distributed evenly about zero, demonstrating an encouraging lack of systematic error in our model. Furthermore, when just the components intended to model the expanding shell were examined (the red, broad Gaussian and the blue Voigt–profile absorption), we found the resulting profile to be in excellent qualitative agreement with the P–Cygni line profiles expected for the transfer of line photons in expanding envelopes (e.g. see Surdej 1979, and references therein).

Table 3.1 summarizes the minimum–χ^2 model parameters. This best–fit model yields a redshift for the central component of the ES1 emission line of $z = 5.190 \pm 0.001$. Most strikingly, to fit the red wing of the emission line, the model demands that the broad emission component be displaced by 320 km s^{-1} from the central emission component. Owing to the strong signal of the Lyα forest at such a high redshift, only a small blue absorption component is required. Still, though of minor amplitude, the absorption component is displaced by fully 360 km s^{-1} from the central emission component. These displacement velocities are in broad agreement with those predicted by Tenorio–Tagle et al. (1999) for the late stages of the evolution of the Lyα profile of a star–forming galaxy. Moreover, though these values somewhat exceed the $\sim 10^2$ km s^{-1} winds typically observed in nearby starbursting dwarf galaxies (e.g. Martin 1998), they compare favorably with the $10^2 - 10^3$ km s^{-1} outflows observed in more powerful local starbursts (Heckman et al. 1990).

We explored a variety of alternative kinematic scenarios for the ES1 emission, most notably by fixing the centroid for the central emission component and arbitrarily sliding the separation between the central component and the broad component over a range of values. These models suffered from worsened χ^2, with broader emission demanded as the displacement velocity was diminished. Figure 3.6 illustrates this trend, implying that even

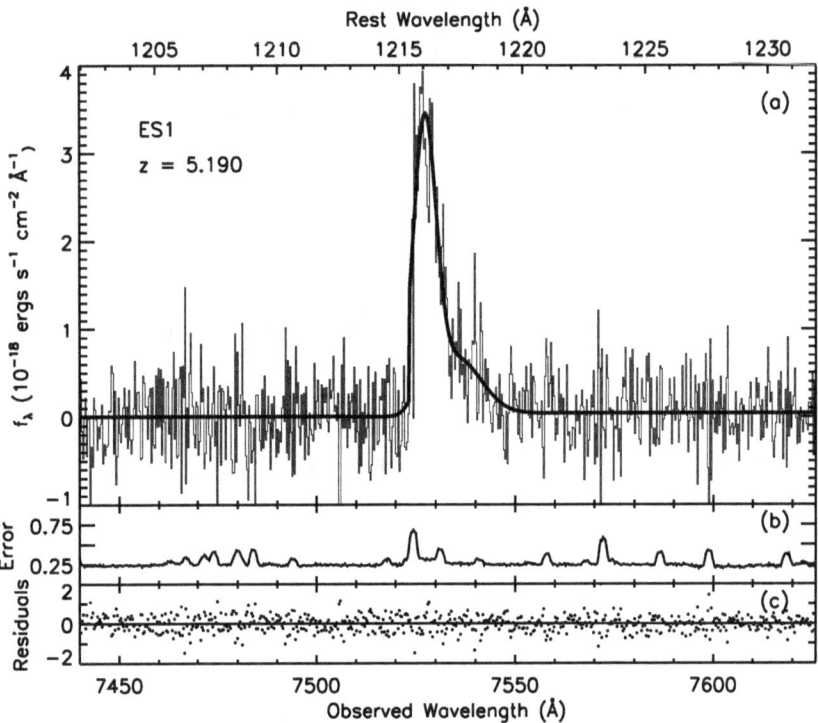

FIG. 3.5.— (a) The minimum–χ^2 fit to the ES1 Lyα emission line. The line profile is the sum of (1) a narrow (280 km s^{-1} FWHM) central Gaussian intended to model recombination in the hot ionized gas of the starburst, (2) a broad (560 km s^{-1} FWHM) Gaussian redshifted by 320 km s^{-1} from the central component intended to model back–scattering off of the far side of an expanding shell, and (3) a broad (800 km s^{-1} FWHM) Voigt absorption component blueshifted by 360 km s^{-1} from the central component intended to model absorption by the near side of the expanding shell. The model spectrum was attenuated by the model of the Lyα forest presented by Madau (1995). (b) The error per pixel in the same flux units and over the same wavelength range. For background–limited observations of faint objects in this region of wavelength space, night–sky emission lines are the dominant source of noise. (c) The model–fit minus the data in the same flux units and over the same wavelength range. The even distribution of the residuals demonstrates a lack of systematic error in the model.

if the minimum-χ^2 model in Figure 3.5 over–estimates the separation between the central component and the broad component of the Lyα emission, the broad component itself can only get broader. That is, no matter what combination of fit–parameters one chooses, there is no escaping a very energetic component to the Lyα emission of ES1. In a similar vein, it is also worthy of note that models which do not include a high–velocity component (e.g. models with a single Gaussian emission component) also suffer from a worsened χ^2 compared to that of the model in Figure 3.5.

3.4   Discussion and Conclusion

The spectral profile of the Lyα emission of ES1 presents evidence for a galaxy–scale outflow with a velocity of $v > 300$ km s^{-1}. Of course, as Heckman et al. (2000) caution, the outflow rate of a galactic wind cannot necessarily be interpreted as the rate at which mass or energy *escapes* into the IGM, since the observable manifestation of an outflow may be produced by material still deep inside the gravitational potential well of the galaxy halo. Nonetheless, the outflow velocity estimated for ES1 far exceeds the escape speed of a nominally low mass ($M < 10^{10} M_\odot$) pregalactic fragment, consistent with the general observation that hot gas can readily escape from dwarf galaxies, though perhaps not from more massive systems (Heckman et al. 2000; Heckman 2000; Martin 1999).

This conclusion bears on a host of cosmological issues surrounding the evolution of galaxies and the IGM at high redshift. Foremost, it suggests that processed material from ES1 will become available to the IGM, potentially providing the enrichment necessary to account for the amount of metals there observed. Indeed, recent observations of C 4 absorption systems along the lines–of–sight to lensed QSOs call for enrichment at increasingly high redshift, beyond even $z > 5$ (e.g. Aguirre et al. 2001; Rauch, Sargent, & Barlow 2001). Additionally, both detailed observations and careful theoretical studies demand a mechanism for pre–heating the material out of which galaxy clusters ultimately collapse and become bound (e.g. Kaiser 1991; Mushotzky & Scharf 1997, and references therein). Here again, galaxy–scale outflows at high redshift are the likely culprit (e.g. Renzini et al. 1993). Finally, galactic winds have proved important in reproducing the faint–end slope of the observed field galaxy luminosity function in semi–analytic models of galaxy formation. Outflows are invoked to suppress star–formation in low–mass dark matter halos, either via the direct escape of gas–phase baryons in the outflow itself (e.g. Somerville & Primack

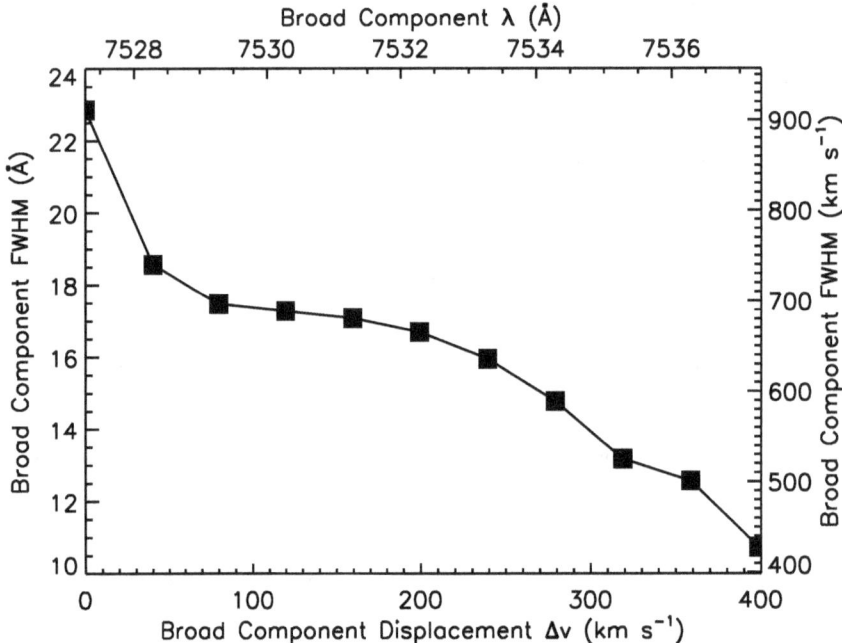

FIG. 3.6.— The relationship between the displacement velocity and the width of the broad emission component. In each case, the centroids for the narrow emission component and the absorption component were fixed at the minimum-χ^2 value; the separation between the narrow emission component and the broad component was set arbitrarily; and all other emission parameters were left unconstrained. When the displacement velocity of the broad component is lowered from its best-fit value, the width of the component increases. Hence, there is no escaping a very energetic component to the Lyα emission of ES1.

1999), or by ram pressure stripping of the gas–phase baryons by energetic winds from neighboring galaxies (Scannapieco & Broadhurst 2001).

As a somewhat speculative conclusion, we now consider the expected correlation between strong galactic outflows and the escape of Lyman continuum radiation from star–forming galaxies. This correlation bears directly on the much–debated physical nature and relative contributions of the sources which comprise the UV background, as a significant contribution by sources other than QSOs is required at high redshift, owing to the rapid decline in the space density of optical and radio–loud quasars at $z > 3$ (Bianchi, Cristiani, & Kim 2001; Madau, Haardt, & Rees 1999).

It is likely that star–forming galaxies fill this niche. From the theoretical standpoint, mechanical energy deposition in the form of supernovae and stellar winds is expected to result in an over–pressured cavity of hot gas inside star–forming galaxies. In galaxies for which the star–formation rate per unit area $\Sigma_* \geq 10^{-1} M_\odot$ yr^{-1} kpc^{-2}, the superbubble ultimately expands and bursts out into the galaxy halo, allowing for the escape of hot gas and facilitating the leak of Lyman continuum photons (Heckman et al. 2000; Tenorio–Tagle et al. 1999). Of course, as the superbubble will expand in the direction of the vertical pressure gradient, the burst is expected to take the form of a weakly collimated, bipolar wind. Hence, the leak of UV radiation may depend sensitively on not only the distribution of neutral gas and dust in the galaxy interstellar medium, but on the inclination of the system. Nonetheless, from the observational standpoint, Steidel, Pettini, & Adelberger (2001, hereafter SPA01) report the detection of significant Lyman continuum emission in a composite spectrum of 29 Lyman–break galaxies at $\langle z \rangle = 3.40 \pm 0.09$, suggesting an escape fraction[4] of UV ionizing photons of $f_{esc} \gtrsim 0.5$.

Given the evidence for a strong outflow in ES1, it would be intriguing to measure the flux of photons below $\lambda = 912$ Å. However, ES1 is very faint even above Lyα; we estimate that it fades to $I_{AB} > 29$ below the Lyman limit. We can do only slightly better at high redshifts with more accessible spectra: even in a composite of four of the highest signal–to–noise Keck spectra of galaxies at $z > 4.5$ collected by Spinrad and collaborators (Figure 3.7), our measurement of the flux of 900 Å photons is consistent with zero at the 1σ level. This result translates to the coarsely constrained flux ratio $f_\nu(1100 \text{ Å})/f_\nu(900 \text{ Å}) =$

---

[4]Here, f_{esc} is the fraction of emitted 900 Å photons that escapes the galaxy without being absorbed by interstellar material, normalized by the fraction of emitted 1500 Å photons which similarly escapes. As SPA01 point out, this definition differs from definitions encountered elsewhere, which typically consider only the fraction of emitted 900 Å photons which escapes (e.g. Bianchi et al. 2001; Heckman et al. 2001; Hurwitz, Jelinsky, & Dixon 1997; Leitherer et al. 1995).

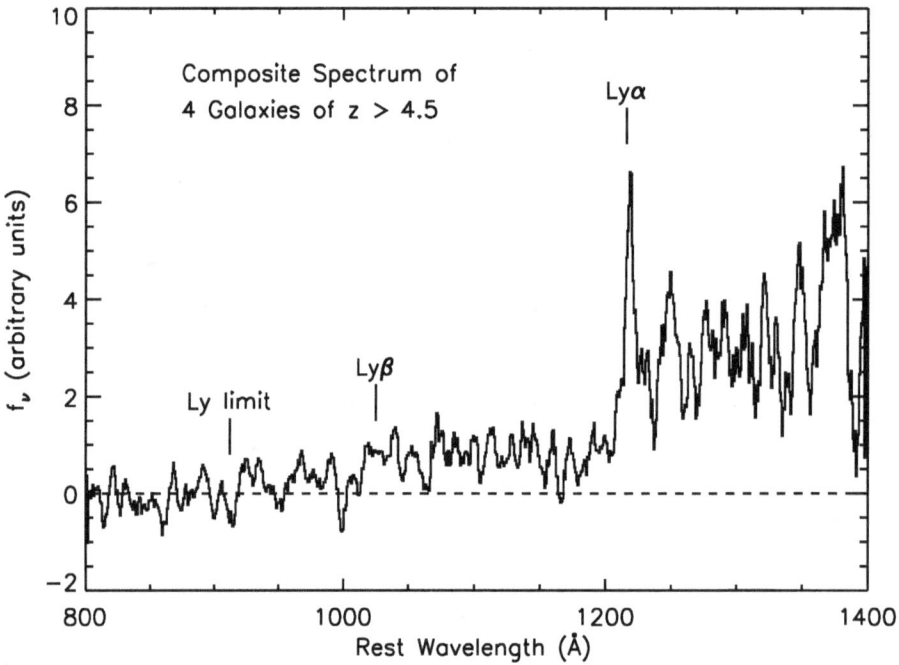

FIG. 3.7.— Composite spectrum of four galaxies at $z > 4.5$: RD581 with $z = 4.89$, HDF 3–951 with $z = 5.34$ (Spinrad et al. 1998), HDF 4–625 with $z = 4.58$, and HDF 4–439 with $z = 4.54$ (both Stern & Spinrad 1999). Following Steidel et al. (2001), the composite was constructed by shifting each flux–calibrated, one–dimensional spectrum to the rest frame, scaling to a common median, and then combining with a simple algorithm which rejected 1σ outliers at each pixel. This rejection scheme removed very nearly one point per pixel. The spectrum has been boxcar–smoothed by one resolution element.

16.7±51.9 (1σ uncertainty). To convert this value to the more useful $f_\nu(1500\ \text{Å})/f_\nu(900\ \text{Å})$ ratio, we adopt an empirical correction factor based on the set of fluxes given in SPA01, yielding an effective $f_\nu(1500\ \text{Å})/f_\nu(900\ \text{Å}) = 31.4 \pm 98.2$. Finally, for the same intrinsic Lyman discontinuity of 3 adopted by SPA1, we find an escape fraction of $f_{esc} \gtrsim 0.1 \pm 0.3$. As we did not correct our initial $f_\nu(1100\ \text{Å})/f_\nu(900\ \text{Å})$ ratio for the opacity of the IGM, this value represents a lower limit. Evidently, to satisfactorily constrain the correlation between outflow dynamics and the escape of UV ionizing photons in an individual high redshift galaxy, we will require spectroscopy of a lensed, blue candidate system viewed along the outflow direction.

Acknowledgements

We are humbly indebted to E. Scannapieco and J. Walters for making generous and substantial contributions to this work, and for providing prodigious comic relief. We are grateful to A. Barger for making public optical images of the HDF and its Flanking Fields of which we have made extensive use; to C. Steidel for providing the target which resulted in this fortuitous detection; to M. Hunt for unwittingly providing software which aided in the reduction of the echelle data; to M. Dickinson for acting as the steward of the HDF and for providing the North West Flanking Field mosaic included in this work; to the anonymous referee for providing careful, useful commentary; and to the expert staff of the Keck Observatory for their invaluable assistance in making the observation. The work of SD was supported by IGPP–LLNL University Collaborative Research Program grant #02–AP–015. HS gratefully acknowledges NSF grant AST 95–28536 for supporting much of the research presented herein. AD acknowledges partial support from NASA HF–01089.01–97A and from NOAO. NOAO is operated by AURA, Inc., under cooperative agreement with the NSF. The work of DS was carried out at the Jet Propulsion Laboratory, California Institute of Technology, under contract with NASA. The work of SD, WvB, WdV, and MR was performed under the auspices of the U.S. Department of Energy, National Nuclear Security Administration by the University of California, Lawrence Livermore National Laboratory under contract No. W–7405–Eng–48. This work made use of NASA's Astrophysics Data System Abstract Service.

TABLE 3.1

ES1 Lyα Line Model Parameters

Component	λ (Å)	Peak Amplitude (10^{-18} ergs s^{-1} cm^{-2} Å^{-1})	FWHM (Å / km s^{-1})	Δv[†] (km s^{-1})
Narrow Component	7527	3.16	7 / 280	⋯
Broad Component	7535	0.62	14 / 560	320
Absorption Component	7518	−0.06	20 / 800	−360

[†] Displacement velocity relative to the central, narrow emission component. Positive velocity is a redshift; negative velocity is a blueshift.

TABLE 3.2

ES1 Ground–Based Photometry

Band	Magnitude (Vega–based, 1″.5 aperture)
V	> 26.1[†]
R	> 27.5[†]
I	25.1 ± 0.1[‡]
z	> 25.2[†]

[†]3σ limiting magnitude.

[‡]Add ∼ 0.4 magnitudes to this Vega–based magnitude for comparison to the AB isophotal magnitude cited in § 3.2.

Chapter 4

Optical and Near–Infrared Spectroscopy of a High–Redshift, Hard X–ray Emitting Spiral Galaxy

A version of this chapter was previously published in *The Astronomical Journal* (Dawson, S., McCrady, N., Stern, D., Eckart, M., Spinrad, H., Liu, M., & Graham, J. 2003, AJ, 125, 1236). Reproduced by permission of the AAS.

Abstract

We present optical and near–infrared Keck spectroscopy of CXOHDFN J123635.6+621424 (hereafter HDFX28), a hard X–ray source at a redshift of $z = 2.011$ in the flanking fields of the Hubble Deep Field–North (HDF–N). HDFX28 is a red source ($R - K_s = 4.74$) with extended steep–spectrum ($\alpha_{1.4\ \mathrm{GHz}}^{8.4\ \mathrm{GHz}} > 0.87$) microjansky radio emission and significant emission (441 μJy) at 15 μm. Accordingly, initial investigations prompted the interpretation that HDFX28 is powered by star formation. Subsequent *Chandra* imaging, however, revealed hard ($\Gamma = 0.30$) X–ray emission indicative of absorbed AGN activity, implying that HDFX28 is an obscured, Type II AGN. The optical and near–infrared spectra presented herein corroborate this result; the near–infrared emission lines cannot be powered by star formation alone, and the optical emission lines indicate a buried AGN. HDFX28 is identified with a face–on, moderately late–type spiral galaxy. Multi–wavelength morphological studies of the HDF–N have heretofore revealed no galaxies with any kind of recognizable spiral structure beyond $z > 2$. We present a quantitative analysis of the morphology of HDFX28, and we find the measures of central concentration and asymmetry to be indeed consistent with those expected for a rare high–redshift spiral galaxy.

4.1   Introduction

The origin of the X–ray background (XRB) remains an enduring puzzle for X–ray astronomy. Great progress has been made during the last four years: *ROSAT* surveys successfully resolved ∼80% of the soft XRB (0.5–2 keV) into discrete sources (e.g. Hasinger et al. 1998) and current work with the *Chandra X–ray Observatory* (hereafter, *Chandra*) has successfully resolved a similar fraction of the hard XRB (2–8 keV; Brandt et al. 2001; Giacconi et al. 2001, 2002; Hornschemeier et al. 2001; Rosati et al. 2002). Nonetheless, a coherent understanding of the physical and evolutionary properties of the sources which comprise the XRB is only just now emerging. Although much of this population appears to be the nuclei of otherwise normal bright galaxies ($I < 23.5$) or typical active galactic nuclei (AGNs), a significant fraction of the discrete sources is optically faint ($I > 23.5$), and therefore not easily identified (e.g. Alexander et al. 2001; Barger et al. 2002). *Type II quasars*, for instance, are thought to be AGNs viewed edge–on through an obscuring torus (Antonucci 1993) and are deemed an essential component of the XRB–producing population (Moran et al. 2001). However, few well–studied examples of such systems are known at high redshift, and owing in part to their lack of relativistic brightening, they are not easily identified in shallow, large–area surveys (Norman et al. 2002; Stern et al. 2002b).

Type II quasars represent just one of several diverse classes of objects emerging in follow–up work to deep *Chandra* fields (e.g. Hornschemeier et al. 2001; Schreier et al. 2001; Stern et al. 2002a). On the extra–galactic side, we find X–ray–loud composite galaxies typified by starburst or early–type optical spectra which bear no signature of their buried AGN (e.g. Moran et al. 1996; Levenson et al. 2001; Stern et al. 2002a). Additionally, we find X–ray sources whose optical counterparts belong to the class of faint, extremely red objects (EROs), the nature of which has remained uncertain owing to the difficulty in spectroscopic follow–up (e.g. Alexander et al. 2002; Elston et al. 1988, 1989; Hu &

Ridgway 1994; Graham & Dey 1996; Liu et al. 2000; Hornschemeier et al. 2001; Stern et al. 2002a). On the Galactic side, we find late–type dwarfs emitting soft X–rays originating in chromospheric activity (e.g. Hornschemeier et al. 2001), and very low mass binary systems emitting hard X–rays driven by accretion (e.g. Stern et al. 2002a). Amidst the emergence of this menagerie of objects, optical and near–infrared spectroscopic follow–up has become increasingly vital not only to identifying the source population of the XRB, but also to elucidating the physics of X–ray sources in general, and to delineating their evolution with redshift.

In one critical facet of this endeavor is simply to distinguish between objects powered by mass accretion onto supermassive black holes (quasars and other AGNs) and those powered by nuclear fusion in stars (normal and starburst galaxies). To this end, we present optical and near–infrared spectra of CXOHDFN J123635.6+621424 (hereafter HDFX28), a hard X–ray source identified with a face–on spiral galaxy at redshift $z = 2.011$ (Figure 4.1). HDFX28 is fortuitously located in the Hubble Deep Field–North inner west (HDF–N IW) flanking field, and was therefore subject to a vast array of follow–up imaging. As such, HDFX28 was initially identified as an extended microjansky radio source with a comparatively steep spectral index ($S_\nu \propto \nu^{-\alpha}$; $\alpha^{8.4 \text{ GHz}}_{1.4 \text{ GHz}} > 0.87$). Together with its detection by the *Infrared Space Observatory* Camera (ISOCAM) and its pronounced optical spatial extent ($\sim 1\rlap{.}''6$), the radio data for HDFX28 prompted an initial interpretation as a galaxy powered by star formation (e.g. Richards 2000). However, as we discuss below, the detection of HDFX28 as a hard X–ray source in the deep *Chandra* survey of the HDF–N (Hornschemeier et al. 2001; Brandt et al. 2001), corroborated by the spectroscopy presented herein, demonstrates that this galaxy in fact harbors an obscured, Type II AGN.

In addition to confirming its AGN status, the spectroscopy of HDFX28 indicates a surprisingly high redshift for an object with identifiable spiral structure. Dickinson (2000) summarizes the results of morphological studies of the HDF–N by reporting a total lack of even plausible candidates for spiral galaxies at $z > 2$. Prompted by this lack of precedent for high–redshift spirals, we present a quantitative study of central concentration and asymmetry in HDFX28 based on the scheme devised by Abraham et al. (1996) for the analysis of large CCD imaging surveys. With the application of a modest morphological k–correction, we find HDFX28 to have morphological parameters consistent with those derived from catalogs of both artificially redshifted nearby spirals, as well as catalogs of *HST* imaging of spirals out to $z \sim 1$ (Abraham et al. 1996).

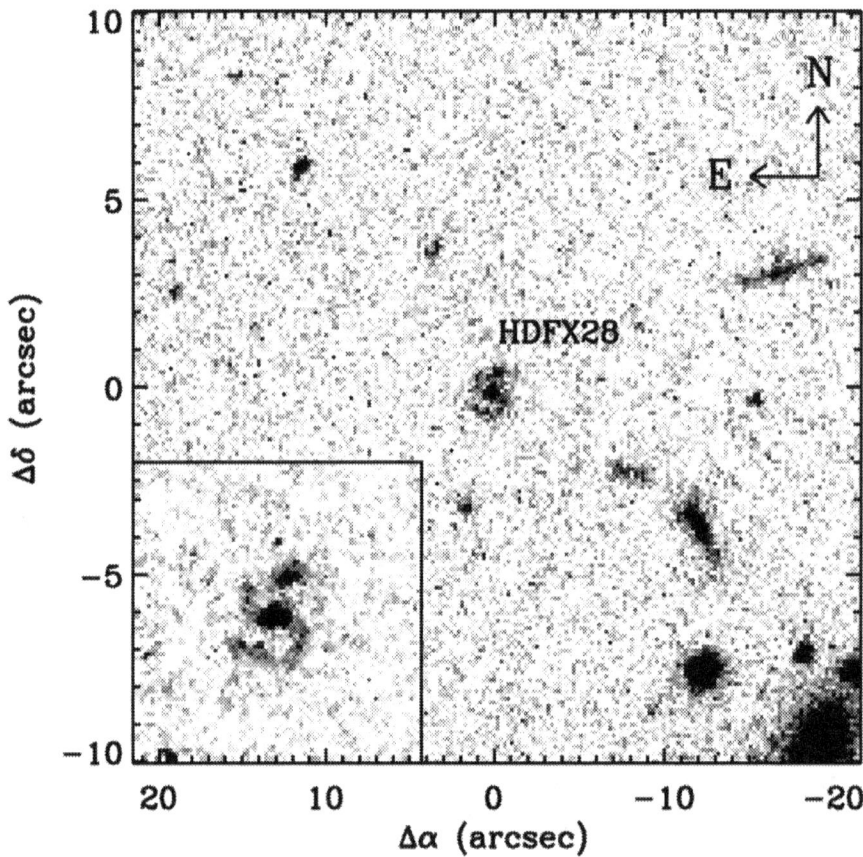

FIG. 4.1.— Central portion of the single–orbit *HST* I_{814} Hubble Deep Field inner west flanking field (Williams et al. 1996), centered on HDFX28 at $\alpha = 12^h36^m35\overset{\prime\prime}{.}6$, $\delta = +62°14'24''$ (J2000). The panel measures $20''$ square and the orientation is indicated. The inset shows a $4''$ square subsection of a two–orbit preliminary image from the first epoch of the Great Observatory Origins Deep Survey (GOODS; Dickinson & Giavalisco 2002) *HST* Treasury Program (L. Moustakas 2002, private communication). The image was taken on UT 2002 Nov 21 with the Advanced Camera for Surveys (ACS; Pavlovsky et al. 2001) and is the sum of 0.5–orbit V_{606} integration, a 0.5–orbit I_{775} integration, and a 1.0–orbit z_{850} integration. With repeat *HST* visits through June 2003, the GOODS program will increase the ACS integration on this field five–fold.

In short, HDFX28 is intriguing both for its membership in the emerging class of X-ray–selected Type II AGN, and for possessing a morphology which is unprecedented at its redshift. We describe the optical and near–infrared spectroscopy of HDFX28 in section §4.2, and we present the results of the spectroscopy and the classification of the source as an obscured, Type II AGN in §4.3. We report on our quantitative analysis of its morphology in §4.4, and we summarize our results in §4.5. Throughout this paper we adopt the currently favored Λ–cosmology of $\Omega_M = 0.35$ and $\Omega_\Lambda = 0.65$, with $H_0 = 65$ km s^{-1} Mpc^{-1} (e.g. Riess et al. 2001). At $z = 2.011$, such a universe is 3.22 Gyr old, the lookback time is 76.9% of the total age of the Universe, and an angular size of $1\rlap{.}{''}0$ corresponds to 8.66 kpc.

4.2   Spectroscopic Observations

4.2.1   Optical Spectroscopy

We obtained the optical spectrum of HDFX28 on UT 2001 February 23 as part of an observing campaign of photometrically–selected high–redshift candidates in the HDF–N and its environs (Dawson et al. 2001). The data were taken with the Low Resolution Imaging Spectrometer (LRIS; Oke et al. 1995) at the Cassegrain focus on the 10m Keck I telescope, after the advent of the LRIS–B spectrograph channel (McCarthy et al. 1998). The red–sensitive LRIS–R camera uses a Tek 2048^2 CCD detector with a pixel scale of $0\rlap{.}{''}212$ pixel^{-1}; the blue–sensitive LRIS–B camera is nearly identical. The data were taken with slitmasks designed to obtain spectra for $\sim$ 15 targets simultaneously through $1\rlap{.}{''}5$ wide slits. For this observation, we used the 400 lines mm^{-1} grating blazed at 8500 Å (1.86 Å pix^{-1} dispersion) in the red channel, and a 300 lines mm^{-1} grism blazed at 5000 Å (2.64 Å pix^{-1} dispersion) in the blue channel. To split the red and blue channels, we used a dichroic with a cutoff at 6800 Å. With this setup, the combined spectrograph channels afforded a spectral coverage of roughly 3200 Å to 1 μm, covering the entire optical window. The total exposure time of 2.75 hours was broken into three exposures of 1500 seconds and three exposures of 1800 seconds; we performed $\sim 3''$ spatial offsets between exposures to facilitate the removal of fringing at long wavelengths. The airmass never exceeded 1.75 during the observations.

We used the IRAF[1] package (Tody 1993) to process the slitmask data, following, where

---

[1] IRAF is distributed by the National Optical Astronomy Observatories, which are operated by the

possible, standard slit spectroscopy procedures. One deviation from standard spectroscopic data reduction was that the blue–channel spectrum was not divided by a flatfield exposure. This omission is a consequence of the fact that the existing internal halogen lamp produces no light shortward of 3800 Å, and also appears to contain prominent UV emission lines. Our experience with attempts at flatfielding LRIS–B in a variety of spectroscopic set–ups indicates that the pixel–to–pixel variations corrected by flatfielding are typically < 4%, and that they have little systematic variation across the CCD; as such, we expect that our flux–calibrated spectra are little affected by this treatment. Remaining aspects of treating the slitmask data were facilitated by a home–grown software package, BOGUS[2], created by D. Stern, A.J. Bunker, and S.A. Stanford.

We extracted the blue–channel and red–channel spectra using the optimal extraction algorithm described in Horne (1986). Wavelength calibrations were performed in the standard fashion using Hg, Ne, Ar, and Kr arc lamps; we employed telluric sky lines to adjust the wavelength zero–point. We performed flux calibrations with longslit observations of standard stars from Massey & Gronwall (1990) taken with the instrument in the same configuration as the multislit observation. However, it should be noted that owing to the constraints of observing with a slitmask, the data were taken at a position angle of $163.1°$, not at the parallactic angle. The final extracted blue–channel optical spectrum is shown in Figure 4.2; the final extracted red–channel optical spectrum is shown in Figure 4.3.

4.2.2   Near–Infrared Spectroscopy

We obtained the near–infrared spectrum of HDFX28 with the 10m Keck II telescope on UT 2001 April 13, using the facility near–infrared spectrometer NIRSPEC (McLean et al. 1998). We employed a $0.''57 \times 42''$ slit to achieve low resolution ($R \sim 1300$) spectra in the wavelength range 1.75–2.17 μm. We obtained four 600 second integrations, with $\sim 5''$ spatial offsets between exposure. The data were dark subtracted, flat–fielded and corrected for cosmic rays and bad pixels in the standard fashion. We sky–subtracted by pairwise subtraction of successive nods along the slit. The curved spectral order was then rectified onto a slit–position/wavelength grid based on a wavelength solution from arc lamp emission lines. As data from the second slit position had significantly lower signal-to-noise, likely the result of temporary seeing degradation or misalignment of the slit, they were

Association of Universities for Research in Astronomy, Inc., under cooperative agreement with the National Science Foundation.

[2]BOGUS is available online at http://zwolfkinder.jpl.nasa.gov/~stern/homepage/bogus.html.

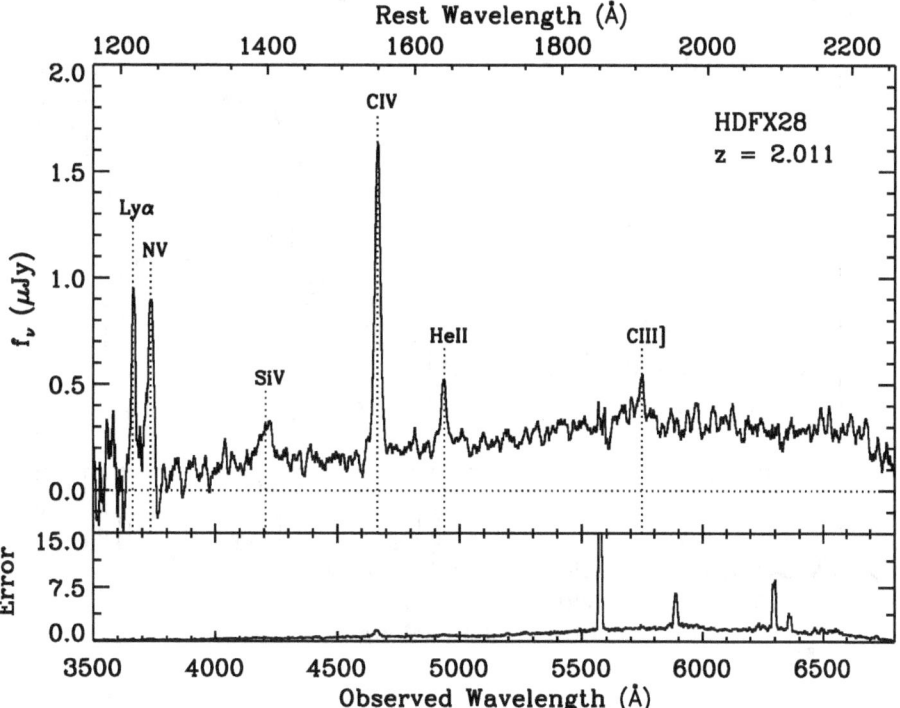

FIG. 4.2.— (Top) Blue channel optical spectrum of HDFX28 obtained with LRIS–B on the Keck I telescope. The spectrum was extracted using the optimal extraction algorithm described in Horne (1986), and was smoothed with a boxcar filter of length equal to one resolution element ($\Delta\lambda \sim 14$ Å at $\lambda = 5000$ Å, based on Gaussian fits to night–sky emission lines). The total integration time was 2.75 hours. (Bottom) The statistical uncertainty per pixel over the same wavelength range and in the same flux units as the object spectrum.

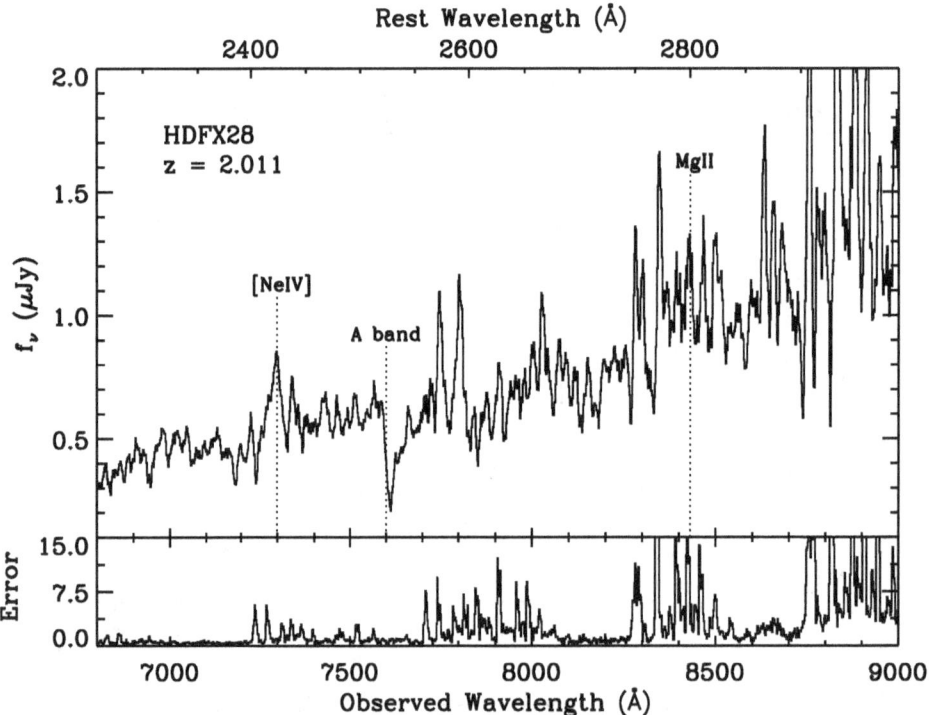

FIG. 4.3.— (Top) Red channel optical spectrum of HDFX28 obtained with LRIS–R on the Keck I telescope. The spectrum was extracted using the optimal extraction algorithm described in Horne (1986), and was smoothed with a boxcar filter of length equal to one resolution element ($\Delta\lambda \sim 11$ Å at $\lambda = 8000$ Å, based on Gaussian fits to night–sky emission lines). The total integration time was 2.75 hours. Note that the unlabeled, narrow spectral features longward of 7800 Å are artifacts due to imperfect subtraction of telluric OH and O_2 night–sky emission lines, and that the absorption feature at 7600 Å is the telluric A–band. (Bottom) The statistical uncertainty per pixel over the same wavelength range and in the same flux units as the object spectrum.

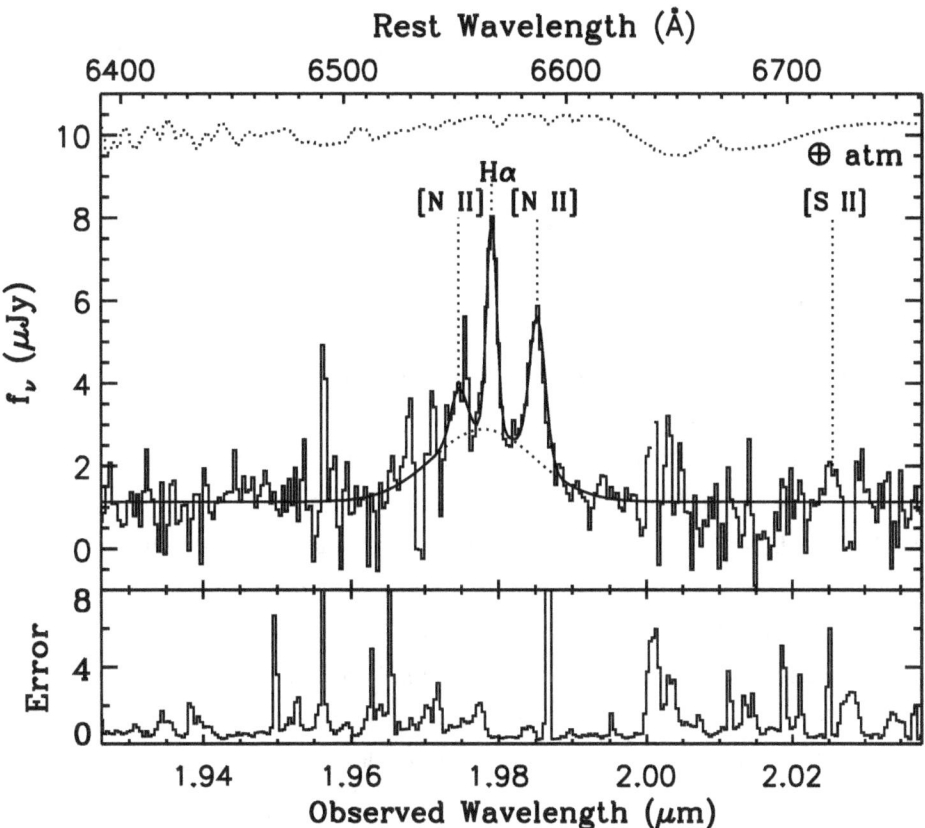

FIG. 4.4.— (Top) Near–infrared spectrum of HDFX28 obtained with NIRSPEC on the Keck II telescope. Four Gaussians were fit to the the emission complex. The total fit (solid curve) consists of [N II] λ6548, [N II] λ6583, and a narrow Hα component, superposed on a broad Hα component (dotted curve). The unresolved [S II] 6716 Å / 6731 Å doublet is barely discernible at ∼ 2.025 μm; it was not included in the fit. The dotted line at the top of the plot shows terrestrial atmospheric absorption, arbitrarily scaled. (Bottom) The calculated error in each wavelength bin in the same units as the object spectrum. The dominant source of error is sky subtraction; the peaks are due to bright sky emission lines which were not well subtracted.

rejected. The total integration time for the near–infrared spectrum is thus 1800 seconds.

The galaxy spectrum was extracted using a Gaussian weighting function which was matched to the wavelength–integrated profile. To correct for atmospheric absorption, we divided the galaxy spectrum by the spectrum of an A0V calibration star, HD 99966. Both the galaxy and the star were observed at an airmass of ~ 1.3. The resulting spectrum is shown in Figure 4.4.

4.3   HDFX28 as a Type II AGN

4.3.1   Results from the Optical Spectrum

The optical spectrum of HDFX28 shows resolved, moderate–width, high–ionization emission lines typical of AGNs (Figures 4.2 and 4.3). Both the permitted and forbidden lines are well–identified, allowing for unambiguous determination of the redshift. To this end, and to ascertain the fluxes and widths of the emission lines, we made a weighted, single Gaussian Levenberg–Markwardt fit to each isolated line[3], resulting in a redshift of $z = 2.011$. We note that this value deviates somewhat from that presented in a recent spectroscopic catalog of *Chandra* sources in the HDF–N (source 142, $z = 2.00$; Barger et al. 2002). The emission line parameters are cataloged in Table 4.1.

The emission line widths in the optical spectrum present a solid case for the classification of HDFX28 in the overall taxonomy of AGN. The canonical definition of a Seyfert 1 galaxy involves a spectrum with broad permitted lines, typically $\gtrsim 5000$ km s^{-1} FWHM, and comparatively narrow forbidden lines, typically ~ 500 km s^{-1} FWHM (e.g. Osterbrock 1989). The definition of a Seyfert 2, by contrast, involves a spectrum showing permitted and forbidden lines of approximately the same FWHM, typically ~ 500 km s^{-1}. On this account, the rough agreement between the widths of the permitted and forbidden UV lines in HDFX28 calls for classification as a Type II source. Moreover, though the permitted line widths of $\gtrsim 1000$ km s^{-1} slightly exceed those expected for a prototypical Seyfert 2, they still fall far short of permitted line widths observed in Type I sources, or in the broad line regions of Type II sources seen in polarized light (e.g. Vernet et al. 2001). In particular, the width of the He II $\lambda 1640$ line compares favorably with the nine high–redshift radio galaxies (HzRGs) presented in Vernet et al. (2001); HzRGs are perhaps the best–studied

---

[3]In the case of the unresolved Si IV doublet, we fit with two Gaussians with amplitudes constrained by the ratio of the doublet Einstein A–values, 1.02:1; due to the large uncertainty in the fit, however, this line was not used in any of the following analysis.

class of obscured, Type II AGN at the redshift of HDFX28 (e.g. McCarthy 1993; Eales
& Rawlings 1993, 1996; Evans 1998). Furthermore, the line widths of HDFX28 compare
favorably with those reported for Type II AGN elsewhere in the literature: e.g. ~ 900 km
s^{-1} for the infrared–selected Type II quasar IRAS 09104+4109 (Kleinmann et al. 1988);
~ 1000 km s^{-1} for the Type II quasar in the *Chandra* Deep Field South, CDF–S 202
(Norman et al. 2002); and $\gtrsim 1000$ km s^{-1} for the Type II quasar in the deep *Chandra*
Lynx field, CXO52 (Stern et al. 2002b).

The emission line flux ratios for HDFX28, however, may somewhat weaken the case
for classification as a straight Type II source. We tabulate the flux ratios for HDFX28
along with those of two other high–redshift Type II AGNs in Table 4.2, and we plot the
sources in the N V λ1240 / He II λ1640 vs. N V λ1240 / C IV λ1549 plane in Figure 4.5.
Both these ratios are comparatively strong in HDFX28, placing it intermediate between
models for the narrow emission lines of HzRGs and models for QSO broad–line regions
(BLRs), though closer to the QSO BLRs. Moreover, though the location of HDFX28 in
Figure 4.5 compares favorably to that of CDF–S 202, HDFX28 is far stronger in both
flux ratios than CXO52. Of course, the N V λ1240 emission in CXO52 was noted as
exceptionally weak; Stern et al. (2002b) report that its N V λ1240 / C IV λ1549 ratio is
approximately half of what is seen in composite HzRG spectra (e.g. McCarthy 1993; Stern
et al. 1999). Nonetheless, the comparative strength of N V λ1240 in HDFX28 may point
to a classification intermediate between Type I and Type II AGN.

The C IV λ1549 / He II λ1640 ratio for HDFX28 indicates a similar conclusion. Typical values of C IV λ1549 / He II λ1640 for unobscured, Type I objects are ~ 10 in composite UV spectra of Seyfert 1 galaxies (Heckman et al. 1995), and 7–50 in composite quasar
spectra (Boyle 1990; Francis et al. 1991; Vanden Berk et al. 2001). Typical values for
obscured, Type II objects are ~ 1 in composite UV spectra of Seyfert 2 galaxies (Heckman
et al. 1995), and ~ 1.5 in composite spectra of HzRGs (McCarthy 1993; Stern et al. 1999).
On this account, HDFX28 is again intermediate between the Type I and Type II sources,
though it is worth noting that the other two high–redshift objects in Table 4.2 somewhat
echo this trend, and both are nevertheless classified as Type II AGN. Still, particularly in
anticipation of the weak, broad Hα emission described below (§4.3.2), we conclude that the
optical spectrum of HDFX28 favors classification somewhere on the continuum between
Type I and Type II sources, rather than as prototypically Type II.

On a separate note, the Lyα line of HDFX28 is exceptionally weak both in equivalent

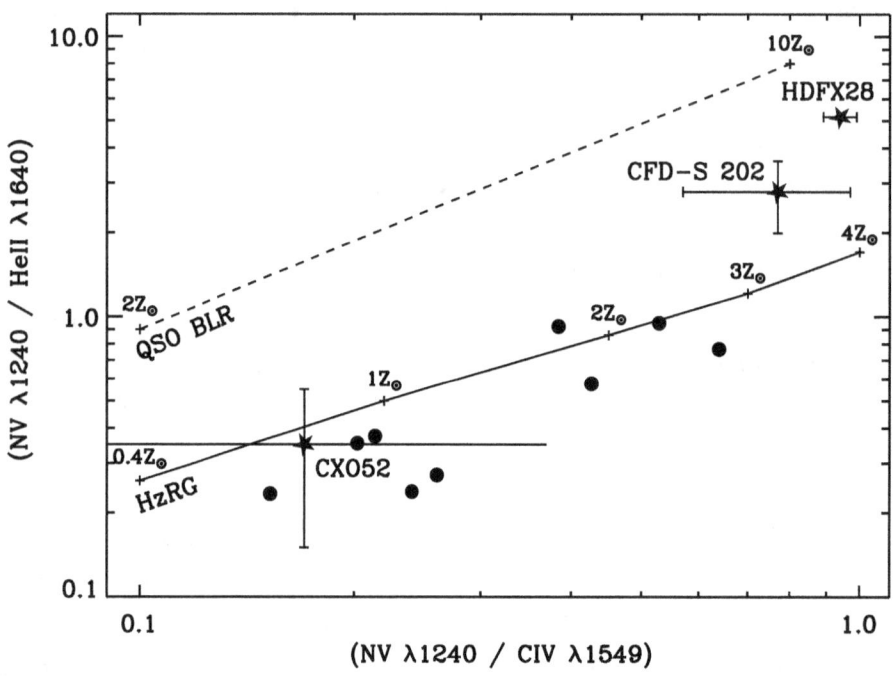

FIG. 4.5.— The N 5 λ1240 / C 4 λ1549 vs. N 5 λ1240 / He 2 λ1640 plane as it appears in Vernet et al. (2001) and Norman et al. (2002), showing HDFX28 along with the Type II quasars CDF–S 202 (Norman et al. 2002) and CXO52 (Stern et al. 2002b). The labeled stars indicate the Type II sources. The circles indicate nine high–redshift radio galaxies (HzRGs) presented by Vernet et al. (2001); the HzRGs are the only class of Type II AGN which have been studied extensively at the redshift of HDFX28. The dashed line represents the locus of a QSO broad–line region (BLR) chemical evolution model with metallicities ranging from 2–10 times solar (Hamann & Ferland 1993). The solid line indicates the locus of the best fit power–law photoionization models for HzRGs with metallicities ranging from 0.4–4 times solar (Vernet et al. 2001). HDFX28 occupies a position intermediate between the two models, and is evidently of high metallicity.

width and in relative flux. For comparison, McCarthy (1993) offers a mean rest–frame equivalent width for HzRGs at $z > 1.5$ of $W_\lambda^{rest}(Ly\alpha) = 295 \pm 188$ Å, and Stern et al. (1999) give a mean rest–frame equivalent width for 17 HzRGs spanning $0.3 < z < 3.6$ of $W_\lambda^{rest}(Ly\alpha) = 75$ Å. Whereas the high–ionization state emission lines of HDFX28 have equivalent widths similar to those reported elsewhere (e.g. ~ 10–10^2 Å for CXO52; Stern et al. 2002b), we find for HDFX28 a meager $W_\lambda^{rest}(Ly\alpha) = 35 \pm 3$ Å. The weakness of $Ly\alpha$ in HDFX28 in relative flux is dramatic both observationally and theoretically. Relative to C IV $\lambda1549$ and N V $\lambda1240$, $Ly\alpha$ emission in both CDF–S 202 and CXO52 exceeds that in HDFX28 by factors ranging from ~ 2 to 10. Furthermore, Ferland & Osterbrock (1986) predict an unreddened $Ly\alpha$ / $H\alpha$ ratio of 16 from a model spectrum of a classical Seyfert 2 galaxy. Anticipating our near–infrared results (below), we report a $Ly\alpha$ / $H\alpha$ ratio for HDFX28 of just 2.4 ± 0.1.

With $E(B-V) = 0.00$ towards the HDF–N (Williams et al. 1996), Galactic extinction is unviable as a culprit for the diminished $Ly\alpha$ flux in HDFX28. Rather, as has been deduced from weak $Ly\alpha$ in several HzRGs (e.g. Eales & Rawlings 1993; Dey et al. 1995), it is likely that dust in HDFX28 is preferentially extinguishing $Ly\alpha$ photons. Since $Ly\alpha$ is a resonant line, $Ly\alpha$ photons are multiply scattered by neutral hydrogen as they traverse the system, resulting in a long path length for dust absorption. Hence, it is possible for $Ly\alpha$ emission to become substantially depressed even if there is little dust in the system, so long as there is sufficient neutral hydrogen.

4.3.2   Results from the Near–Infrared Spectrum

The near–infrared spectrum of HDFX28 shows weak continuum emission and a re-solved emission line complex near 1.98 μm identified as $H\alpha$ plus [N 2] $\lambda\lambda6548, 6583$ Å. As is common in the spectra of Seyfert galaxies, the $H\alpha$ emission consists of two compo-nents: a strong narrow line superposed upon a weak, broad line (e.g. Osterbrock 1989). We therefore fit the emission complex with four Gaussians subject to the following con-straints: (1) the ratio of amplitudes of the [N II] doublet lines must be 2.96:1 as prescribed by the ratio of their Einstein A–values, (2) the [N II] lines must have the same redshift as the narrow $H\alpha$ component, and (3) the two [N II] lines must have identical FWHMs. Again, we made a Levenberg–Markwardt fit to the spectrum, weighting according to the uncertainty of each pixel. The resulting fit is overlaid on the spectrum in Figure 4.4, and the corresponding emission line parameters are cataloged along with the results from the optical spectroscopy in Table 4.1.

Like the optical spectrum discussed above, the near–infrared spectrum indicates a somewhat mixed result for the AGN classification of HDFX28. The widths of the [N II] doublet lines and of the narrow component of the Hα emission are comparable to the narrow lines of a classic Seyfert 2 (e.g. Osterbrock 1989), and they compare favorably with both the forbidden and permitted optical line widths reported for CXO52 in Stern et al. (2002b). However, these results must be mitigated by the presence of weak, broad Hα emission. We find a FWHM of 2500 ± 250 km s^{-1} for this broad Hα component, and a ratio of broad–to–narrow emission of Hα(b)/Hα(n) $= 3.3 \pm 0.3$. For comparison, the broad component to the Hα emission in HDFX28 is far weaker than that of the HzRG MRC 2025–218 (Larkin et al. 2000), with its Hα(b) FWHM of 9300 ± 900 km s^{-1} and broad–to–narrow flux ratio of Hα(b)/Hα(n) $= 7 \pm 2$. Stern et al. (2002b) point out that, though the spectrum of MRC 2025–218 over the range from Lyα to [O III] λ5007 is very similar to classic obscured, Type II AGN observed by other groups (e.g. Eales & Rawlings 1993, 1996; Evans 1998), the presence of broad Hα may indicate that MRC 2025–218 is actually the high–luminosity analog to a Seyfert 1.8, rather than a Seyfert 2. In keeping with the results of the optical spectroscopy, the same claim may therefore be made for HDFX28.

One further diagnostic offered by the near–infrared spectrum of HDFX28 is the ratio of its [N II] λ6583 flux to its Hα(n) flux. Veilleux & Osterbrock (1987) and Osterbrock (1989) present a classification scheme employing these and other optical features to discriminate the narrow lines of AGNs from those of starburst galaxies. The physical distinction exploited in this case is the differing strengths of low–ionization lines such as [N II] λ6583 in each class of source. In the narrow line region of an AGN, these low–ionization lines arise preferentially in an extended zone of partly photoionized hydrogen which results from an ionizing spectrum containing a large fraction of high–energy photons. These photons are absent in the spectrum of OB stars; hence, the strength of the low–ionization lines is diminished in starburst galaxies (e.g. Baldwin et al. 1981; Veilleux & Osterbrock 1987; Osterbrock 1989). Typically, starbursts and H 2 region–like galaxies occupy $-2 \lesssim \log([\text{N II}] \; \lambda6583/\text{H}\alpha(\text{n})) \lesssim -0.3$, while AGN occupy $-0.3 \lesssim \log([\text{N II}] \; \lambda6583/\text{H}\alpha(\text{n})) \lesssim 0.8$ With reference to Table 4.2, we find HDFX28 to fall definitively within the regime of AGNs.

This result is corroborrated by the rest–frame equivalent width of the near–infrared emission complex. Based on an average continuum level of 1.13 ± 0.06 μJy, the total equivalent width of the (Hα + [N II]) feature is $W_\lambda^{\mathrm{rest}} = 150 \pm 40$ Å. By contrast, surveys of normal (i.e. non–AGN) galaxies find an average width in the range of just 20–30 Å (e.g. Kennicutt & Kent 1983), and surveys of starburst galaxies find only $\sim$ 40 Å (e.g. Ravindranath & Prabhu 2001).

4.3.3   Results from the Multi–Wavelength Photometry

The aim of the *Hubble Space Telescope* (*HST*) observations of the HDF–N was to image an otherwise undistinguished field as deeply as reasonably possible (Williams et al. 1996). Indeed, the HDF–N represents the deepest optical images ever taken, providing detections and photometry of stars and field galaxies to $V \sim 30$ with $0\rlap{.}''1$ resolution, and reaching source densities of $\sim 10^6$ deg^{-2} (for a review, see Ferguson et al. 2000). One caveat is that the *HST* images of the HDF–N are rather small, covering only $\sim$ 5 arcmin^2. Hence, to facilitate ground–based follow–up observations, the deep imaging program was augmented with short, 1–2 orbit images of eight fields immediately adjacent to the primary field. These flanking field observations were made exclusively with the WFPC2 I_{814} filter (Williams et al. 1996, Table 2).

HDFX28 is located 1.6′ west and 1.4′ north of the pointing center of the primary HDF–N in the inner west flanking field. Its optical counterpart gives the impression of a moderately late–type face–on spiral galaxy measuring roughly $1\rlap{.}''6$ (14 kpc) in diameter (Figure 4.1). For their program of associating *Infrared Space Observatory* (*ISO*) detections with optical sources in and adjacent to the HDF–N, Mann et al. (1997) constructed an I_{814} catalog of the flanking fields; they give $I_{814} = 23.46$ for HDFX28.

Owing to its location in the HDF–N flanking fields, HDFX28 has been inadvertently subject to a panoply of follow–up imaging (see Table 4.3). The galaxy was first reported as a weak radio source (8.15 μJy at 8.5 GHz; 87.8 μJy at 1.4 GHz) in the sensitive radio surveys of Richards et al. (1998) and Richards (2000), respectively. These results yield a comparatively steep radio spectral index ($S_\nu \propto \nu^{-\alpha}$; $\alpha_{1.4\,\mathrm{GHz}}^{8.4\,\mathrm{GHz}} > 0.87$), with radio emission extending across $2\rlap{.}''8$. In general, microjansky radio emission from disk galaxies can result from either star formation (e.g. from free–free emission originating in H 2 regions) or from AGN activity connected with a central engine. Richards (2000) argued that (1) in the case of a central AGN powering a weak ($P < 10^{25}$ W Hz^{-1}) radio source, the bulk of the radio emission is confined to the nuclear region and is therefore characterized by sub–arcsecond angular scales, and (2) such small scales result in a high opacity to synchrotron self–absorption, yielding flat or inverted spectral indices typically in the range $-0.5 < \alpha < 0.5$. HDFX28 is indeed a weak radio source, with $L_{1.4\,\mathrm{GHz}}^{\mathrm{rest}} \sim 2 \times 10^{24}$ W Hz^{-1}. Operating under the assumption that HDFX28 has a redshift in the range $0.2 < z < 1$ (as was inferred from the spatial extent of HDFX28 in the flanking field image, and was consistent with the bulk of the microjansky radio sources in deep VLA surveys), Richards (2000) would have estimated even less radio power. Hence, the origin of the radio emission in HDFX28 was taken to be extended star–forming regions.

This conclusion was ostensibly borne out by the ISOCAM detection of HDFX28 (Aussel et al. 1999). If HDFX28 were a starburst galaxy at moderate–to–low redshift, then the ISOCAM 15 μm filter (LW3) would sample rest wavelengths from roughly 6 μm to 12 μm. The mid–infrared emission could therefore be plausibly attributed to the unidentified infrared bands (UIB) and to the hot, 200 K dust which typically dominates the spectral energy distribution of starbursts over those wavelengths (Aussel et al. 1999). Together, the radio and mid–infrared data therefore appeared to paint a coherent picture of HDFX28 as a source with star formation in its disk as its underlying emission mechanism. This conclusion was consistent with the general observation from deep VLA surveys that the bulk of the radio population at the microjansky level consists of starforming disks, with fewer than 20% of the radio sources associated with early–type galaxies or quasars (Fomalont 1996; Richards et al. 1998).

Subsequent ground–based optical and near–infrared imaging found HDFX28 to be comparatively red, falling just short of the conventional definition of EROs (e.g. $R - K_s > 5.0$; Hornschemeier et al. 2001, and $I - K_s > 4$; Stern et al. 2002a). Hogg et al. (2000) give $R - K_s = 4.74$ for HDFX28, and Barger et al. (2000) add $I - K_s = 3.89$. In contrast to the conclusion drawn from the radio and IR imaging discussed above, and in anticipation of the X–ray data discussed below, it is notable that HDFX28 corroborates the trend that X–ray sources at modest R band magnitudes tend to be redder than typical field galaxies, and that in the X–ray population there appears to be an excess of sources with $4 < R - K_s \lesssim 5$ (see Hornschemeier et al. 2001, Figure 7). It is yet more notable that Hasinger (1999) reports that all X–ray counterparts with $(R - K') > 4.5$ in the ROSAT Ultra Deep HRI Survey are either members of high–redshift clusters or are obscured AGNs.

On that note, the 1 Ms *Chandra* survey of the HDF–N and its environs yielded a soft X–ray flux of 0.28×10^{-15} ergs cm^{-2} s^{-1} and a hard X–ray flux of 2.82×10^{-15} ergs cm^{-2} s^{-1} for HDFX28 (Brandt et al. 2001). Together with the optical imaging, these results correspond to an X–ray to optical flux ratio[4] of $\log(f_X/f_R) = -0.65$ in the soft band and $\log(f_X/f_R) = 0.35$ in the hard band. As noted by Hornschemeier et al. (2001) and Stern et al. (2002a), the majority of X–ray sources in shallow surveys fall within $-1 < \log(f_X/f_R) < 1$, as this range is typical of local AGN. In particular, these values compare very favorably to the Type II QSO CDF–S 202 (Norman et al. 2002), with its soft band ratio of $\log(f_X/f_R) = -0.61$ and its hard band ratio of $\log(f_X/f_R) = 0.29$.

The X–ray data indicate that HDFX28 is a comparatively hard source. Following the nomenclature of Stern et al. (2002a), the hardness ratio for HDFX28 is $HR = 0.24 \pm 0.10$, comparing favorably to $HR = 0.07 \pm 0.13$ for the Type II quasar CXO52. Brandt et al. (2001) report an X–ray band ratio for HDFX28 of $1.66^{+0.37}_{-0.30}$ and a corresponding estimate of the photon index[5] of $\Gamma = 0.30$. These results show HDFX28 to be distinctly hard for its soft band count rate compared to the total samples in both the *Chandra* survey of the HDF–N (Brandt et al. 2001) and the *Chandra* survey of the Lynx field (Stern et al. 2002a). Moreover, this photon index is quite unlike the steep $\Gamma \sim 1.7 - 2.0$ indices typical of unobscured AGNs (e.g. Nandra & Pounds 1994). By assuming that HDFX28 has an intrinsic power law spectrum with $\Gamma = 1.8$ such that the observed band ratio is due to obscuration at the source, and by adopting a Galactic absorption column density in the direction of the HDF–N of $N_{\rm H} = 1.7 \times 10^{20}$ cm^{-2} (Williams et al. 1996), we estimate the hydrogen column density at the source to be $N_{\rm H} \sim 1.5 \times 10^{23}$ cm^{-2}. This value places HDFX28 very near to the median $N_{\rm H}$ of the sample of 73 nearby Seyfert II galaxies compiled by Bassani et al. (1999), and it implies an unobscured full band rest–frame luminosity of 1.1×10^{44} erg s^{-1}, which is well within the quasar regime. In short, each of these results point to significant soft X–ray absorption by intervening material within HDFX28. Hence, as first suggested by Hornschemeier et al. (2001) and as confirmed by the optical and near–infrared spectroscopy presented herein, the X–ray data show HDFX28 to be an obscured, Type II AGN.

---

[4]Hornschemeier et al. (2001) use the Kron–Cousins R filter transmission function to derive the X–ray to optical flux ratio: $\log(f_X/f_R) = \log f_X + 5.50 + R/2.5$.

[5]The photon index Γ is derived from a power law model for the X–ray spectrum: $N = AE^{-\Gamma}$, where N is the number of photons s^{-1} cm^{-2} keV^{-1} and A is a normalization constant (e.g. Hornschemeier et al. 2001).

4.4   HDFX28 as a High–Redshift Spiral Galaxy

It is surprising to find identifiable spiral structure at the early time indicated by the redshift of HDFX28. Careful, multi–wavelength morphological studies of the HDF–N reveal no galaxies with any kind of recognizable spiral structure at $z > 2$ (Dickinson 2000). To wit, the redshift distribution of a sample of 52 late–type spiral and irregular galaxies complete to $K < 20.47$ shows a dramatic cut–off at $z \sim 1.4$, with only two galaxies in the sample exceeding this limit (Rodighiero et al. 2000). Similarly, a combined photometric redshift / morphological data set complete to $I < 26.0$ shows a sharp drop in the spiral galaxy distribution at $z > 1.5$ (Driver et al. 1998). Specifically, in the $22 < I_{AB} < 23$ magnitude bin, there are no spiral galaxies beyond $z > 1.5$; in the $23 < I_{AB} < 24$ magnitude bin, there is only one.

Abraham et al. (1994, 1996) cautioned that visual morphological classifications of late–type galaxies fainter than $I = 21$ are somewhat subjective, particularly for distant systems with small image sizes. This difficulty is most pernicious for *very* late spirals (morphological type $T > 7$), and for merging systems and peculiar galaxies. Nonetheless, especially when combined with the lack of precedent for spiral galaxies at the redshift of HDFX28, this caveat prompted us to bolster our qualitative, visual classification with a quantitative, objective classification.

To this end, we employed a morphological classification scheme devised by Abraham et al. (1996) for analysis of the current generation of large CCD imaging surveys, and modified by Kuchinski et al. (2001) for patchy, low signal–to–noise data. The classification scheme is a two–part system which uses quantitative measurements of the galaxy central concentration and asymmetry to distinguish three morphological bins: E/S0 galaxies, spiral galaxies, and irregular or peculiar systems. The concentration index (C) is the ratio of the light emitted from a central region of the galaxy (usually $R < 0.3R_{max}$, where R_{max} is the radius of an elliptical aperture centered on the galaxy) to the light emitted from the galaxy as a whole. The asymmetry index (A) is a measure of the $180°$ rotational symmetry of the galaxy, measured by rotating the galaxy image about the central pixel and subtracting the rotated image from the original image. In essence, galaxies with high degrees of central concentration and symmetry have regular, ordered appearances, roughly corresponding to early to mid–Hubble types. Galaxies with low central concentration and large asymmetry have irregular or peculiar morphologies, corresponding to late to irregular Hubble types.

We calculated C and A for HDFX28 using the definitions given by equations (1) and (3) in Kuchinski et al. (2001). It has been shown that these indices are sensitive to the definition of the center of the galaxy image and to the aperture in which the indices are measured (Kuchinski et al. 2001, and references therein). As such, we determined the central pixel by first smoothing the galaxy image with a Gaussian kernel of $\sigma = 1$ pixel and then taking the location of the maximum pixel as the galaxy center. We defined the aperture by setting a threshold at $1.0\sigma_{sky}$ and then defining an ellipse based on the intensity–weighted moments of the resulting image. Both these methods have their precedent in a significant body of similar work (e.g. Abraham et al. 1996; Teplitz et al. 1998); we effected the aperture definition with the source extraction software package SExtractor (Bertin & Arnouts 1996). For the concentration index, we found $\log C = -0.43 \pm 0.03$; for the asymmetry index, we found $\log A = -0.37 \pm 0.07$. As we discuss below, these objective results are indeed consistent with our qualitative classification of HDFX28 as a spiral galaxy.

We estimate the uncertainty in C and A by considering two independent sources of error: the statistical error due to Poisson noise entering into the calculations (σ_P), and the variance introduced by calculating C and A in apertures extending to different limiting surface brightnesses (σ_S). For C, the uncertainty due to noise was determined by propagating the Poisson noise per pixel through the calculation in the standard fashion; we found $\sigma_{P,C} = 0.021$. The uncertainty inherent in using apertures defined to different limiting surface brightnesses was quantified by calculating C for 11 apertures of decreasing size determined by running SExtractor with detection thresholds spanning $0.5\,\sigma_{sky}$ to $1.5\,\sigma_{sky}$. The standard deviation of these measurements was $\sigma_{S,C} = 0.026$. Adding these uncertainties in quadrature yielded our total uncertainty estimate of $\sigma_C = 0.03$.

To calculate A, the absolute value is taken of the difference between the original image and the rotated image, resulting in sky noise which systematically contributes only positive values. As in Abraham et al. (1996), we corrected for this effect by subtracting from A the measured asymmetry of many (10^2) blank patches of sky with apertures equal to that enclosing the galaxy. At the same time, we estimated the Poisson error by measuring the distribution of the asymmetry indices of these sky–only apertures. This process yielded $\sigma_{P,A} = 0.058$. The uncertainty due to using apertures defined by different limits was determined for A exactly as it was determined for C with the result $\sigma_{S,A} = 0.037$. Together, these considerations yielded a total estimated error of $\sigma_A = 0.07$.

To interpret our results in terms of Hubble types, we compare HDFX28 to two large reference samples for which the concentration and asymmetry indices have been calculated: (1) the Frei et al. (1996) catalog of nearby galaxies artificially redshifted to $z = 0.3, 0.5$, and 0.7 (Abraham et al. 1996, figure 2), and (2) the catalog of galaxies imaged in the *HST* Medium Deep Survey (MDS), which are expected to have a redshift distribution spanning $0 < z < 1.0$ with a peak at $z = 0.6$ (Abraham et al. 1996, figures 5 and 6). Our initial impression is that the central concentration index of HDFX28 is very typical of spiral galaxies, but that the asymmetry index straddles the border in A between the spirals (low A) and the peculiars (high A). However, as HDFX28 is at a higher redshift than the most distant objects in either of these samples, proper interpretation of its morphological indices requires that we first consider the effects of cosmological distances on morphology.

Three effects complicate the issue of morphology for galaxies at high–redshift: band-shifting, surface brightness dimming, and the loss of spatial resolution. Based on a careful study of *in situ* ultraviolet and optical imaging of 32 local galaxies, Kuchinski et al. (2001) report that bandshifting is the dominant effect. As such, the general trend is for C to decrease and A to increase as one proceeds to higher redshifts. Among other effects, at shorter rest wavelengths apparent morphology becomes dominated by localized star formation, thereby diminishing the effect of an optical bulge (if any) on C and increasing the patchiness measured by A. As for HDFX28, we note that the effect of bandshifting on C is less pronounced in later spiral galaxies that lack dominant bulges to begin with; the Sbc–Sd spirals in the Kuchinski et al. (2001) sample show an average move of $\Delta C \sim 0.1$, where $\Delta C = C_{\mathrm{OPT}} - C_{\mathrm{FUV}}$. Hence, we would expect little change in C if we were able to observe HDFX28 closer to its rest–frame optical, or if it were located at the modest redshifts of the catalogs described above.

The opposite is true of A. Star formation in the disk of a spiral is UV bright, producing large measures of asymmetry in the UV even though the galaxy may appear symmetric in the optical. Kuchinski et al. (2001) found that ΔA between UV images and optical images of a galaxy can be as large as -0.7 (again in the sense of $\Delta A = A_{\mathrm{OPT}} - A_{\mathrm{FUV}}$), though $\Delta A \sim -0.3$ is more typical of later spirals with moderate values of A_{FUV} (Kuchinski et al. 2001, figure 4).

As for surface brightness dimming and the loss of spatial resolution, Kuchinski et al. (2001) report that C is in most cases robust to both effects out to $z \sim 3$ ($\Delta C \lesssim 5\%$), while the effect on A is simply to increase its scatter ($\Delta A \lesssim 12\%$). Consequently, while we interpret bandshifting as resulting in a systematic shift in C and A, we interpret the scatter introduced by surface brightness dimming and the loss of spatial resolution as an increase in their error bars. Hence, the morphological k–correction necessary to properly compare HDFX28 to the artificially redshifted Frei catalog and to the galaxies of the MDS amounts to an increase of ~ 0.1 in concentration index to $\log C \sim -0.33$, with a corresponding increase in error bar to $\sigma_C = 0.04$. Similarly, the asymmetry index must be shifted down by ~ -0.3 to $\log A \sim -0.90$, with a corresponding increase in error bar to $\sigma_A = 0.09$. Once applied, these consideration show HDFX28 to fall definitively within the spiral galaxies in the $\log C - \log A$ distribution of both samples (Abraham et al. 1996). Therefore, we judge the morphology of HDFX28 to be consistent with that of a rare, high–redshift spiral.

The inapplicability of the classical Hubble tuning fork to the galaxy population at $z \gtrsim 0.5$ has been well–documented, and recent morphological studies have shown the dearth of spirals at high redshift to be a genuine change in the galaxy population — not merely a function cosmological distance effects on morphological classification (e.g. van den Bergh et al. 2002). Possible scenarios posited to explain this effect include the destruction of early–time disks from without by mergers or from within by strong, starburst–driven galactic winds, or perhaps stellar feedback in early disks suppresses the cooling of gas before $z \sim 1$, preventing global dynamical instabilities from initiating the formation of spiral structure (van den Bergh 2002). In any case, this single detection of an object at a redshift for which spirals are not expected is certainly not a challenge to widely–accepted hierarchical evolutionary scenarios, which have otherwise been successful at predicting the results of deep galaxy surveys (e.g. Kauffmann et al. 1993; Baugh et al. 1998). It may be the case that HDFX28 is simply a rare example of an early–time disk which escaped destruction, for instance, by a merger event.

4.5   Conclusion

We have reported on two aspects of the high–redshift, hard X–ray emitting spiral galaxy HDFX28: (1) its classification as a Type II AGN, a population recently attracting renewed interest due to deep X–ray surveys, and for which few *HST* images are available, and (2) its unprecedented redshift for a galaxy with spiral morphology. As for HDFX28 as a Type II AGN, the canonical wisdom regarding weak, extended radio sources with spectral indices steeper than $\alpha_{1.4\ \mathrm{GHz}}^{8.4\ \mathrm{GHz}} > 0.5$ dictates that such sources are driven by star formation. Nonetheless, the combined weight of evidence from X–ray, optical, and near–infrared observations of HDFX28 indicates the presence of obscured AGN activity. It is instructive to note that when re–interpreted in light of the spectroscopic redshift, even the mid–infrared data for HDFX28 corroborates this result. At $z = 2.011$, the ISOCAM LW3 filter samples rest wavelengths spanning only 4 μm to 5 μm. Here, the contribution to the mid–IR spectral energy distribution made by UIB emission and by dust at 200 K is severely attenuated (see Aussel et al. 1999, Figure 1). Hence, the ISOCAM detection of this source is far more plausibly explained by the hot, $\sim 10^3$ K dust found in the central region of an AGN (e.g. see Aussel et al. 1998) than it is by star formation alone.

As to the precise nature of the central engine in HDFX28, we conclude from the comparatively narrow emission lines in the spectroscopy and from the heavy obscuration evident in the X–ray data that HDFX28 is far more like an obscured Type II system than an unobscured Type I system. Though this conclusion is slightly at odds with the presence of weak, broad Hα emission, all remaining aspects of the source are entirely consonant with observations of other Type II AGN at moderate–to–high redshifts (e.g. Kleinmann et al. 1988; Norman et al. 2002; Stern et al. 2002b) and with HzRGs (e.g. Larkin et al. 2000; McCarthy 1993; Stern et al. 1999; Vernet et al. 2001). Norman et al. (2002) describe a very similar situation in which their source CDF–S 202 shows both the narrow (~ 1000 km s^{-1}) emission lines in its optical spectrum and the heavy obscuration in its X–ray emission typical of a Type II system, but also shows emission line flux ratios intermediate between Type I and Type II systems. As noted by Stern et al. (2002b), it is conceivable that longer–wavelength spectra of CDF–S 202 and other sources like it would also reveal broad Hα, though they in every other way give evidence of the heavy obscuration considered to be emblematic of Type II AGN.

Separately, the spectroscopy presented herein shows HDFX28 to be at an unprecedented redshift for a galaxy with identifiably spiral structure. Nevertheless, with the application of a modest morphological k–correction, our quantitative analysis of its central concentration and asymmetry is consistent with the interpretation that HDFX28 is a rare example of a high–redshift spiral galaxy. Owing to its proximity to the HDF–N, HDFX28 will be subject to deep, space–based B, V, i, and z imaging with the Advanced Camera for Surveys as part of the upcoming GOODS HST Treasury Program (M. Giavalisco, PI), as well as to infrared imaging at $\lambda > 3$ μm with the Infrared Array Camera as part of the GOODS $SIRTF$ Legacy project (M. Dickinson, PI). At a minimum, the availability of multi–wavelength imaging will provide a powerful additional lever arm on the issue of the morphology of HDFX28 (e.g. Conselice 1997; Conselice et al. 2000). As such, we eagerly look forward to these expansive datasets.

Acknowledgements

We are grateful to Leonidas Moustakas, Mark Dickinson, Mauro Giavalisco, and the GOODS team for kindly providing the preliminary ACS imaging of the HDF inner west flanking field. In addition, we are indebted to the expert staff of the Keck Observatory, without whom this work would not have been possible, and to J. G. Cohen and C. C. Steidel for supporting LRIS–R and LRIS–B, respectively. We gratefully acknowledge the careful reading and useful commentary of the anonymous referee, by which this work substantially benefited. Finally, the authors wish to acknowledge the significant cultural role that the summit of Mauna Kea plays within the indigenous Hawaiian community. We are fortunate to have the opportunity to conduct observations from this mountain. The work of SD was supported by IGPP–LLNL University Collaborative Research Program grant #02–AP–015, and was performed under the auspices of the U.S. Department of Energy, National Nuclear Security Administration by the University of California, Lawrence Livermore National Laboratory under contract No. W–7405–Eng–48. The work of DS was carried out at the Jet Propulsion Laboratory, California Institute of Technology, under contract with NASA. HS gratefully acknowledges NSF grant AST 95–28536 for supporting much of the research presented herein. ML is grateful for research support from the Beatrice Watson Parrent Fellowship at the University of Hawai'i. This work made use of NASA's Astrophysics Data System Abstract Service.

Table 4.1

Emission–Line Measurements of HDFX28

Line	$\lambda_{\rm obs}$ (Å)	Redshift	Flux $(10^{-17}$ ergs cm^{-2} s$^{-1})$	FWHM[†] (km s^{-1})	$W_\lambda^{\rm rest}$ (Å)
Lyα	3664.4 ± 0.7	2.0135 ± 0.0006	1.68 ± 0.08	1270 ± 30	35 ± 3
N V λ1240	3733.5 ± 0.7	2.0109 ± 0.0006	2.3 ± 0.1	2110 ± 30	50 ± 3
C IV λ1549	4665.1 ± 0.7	2.0117 ± 0.0005	2.47 ± 0.08	1300 ± 10	60 ± 3
He II λ1640	4935.8 ± 0.7	2.0096 ± 0.0004	0.45 ± 0.08	1400 ± 60	13 ± 3
C III] λ1909	5745.9 ± 0.8	2.0099 ± 0.0004	0.19 ± 0.08	900 ± 130	7 ± 3
[Ne IV] λ2424	7292.9 ± 0.7	2.086 ± 0.0003	0.3 ± 0.1	1470 ± 30	9 ± 7
[N II] λ6548	19835.2 ± 0.7	2.0155 ± 0.0001	0.24 ± 0.08	380 ± 30	9 ± 11
Hα(n)	19790.5 ± 0.7	2.0155 ± 0.0001	0.7 ± 0.1	240 ± 30	26 ± 9
Hα(b)	19780 ± 8	2.014 ± 0.001	2.3 ± 0.3	2500 ± 250	90 ± 30
[N II] λ6583	19897.3 ± 0.7	2.0155 ± 0.0001	0.72 ± 0.09	380 ± 30	30 ± 12

[†]The line widths have been deconvolved according to the relation FWHM(obs)^2 = FWHM(instr)^2 + FWHM(inher)^2, where FWHM(obs) is the observed line width, FWHM(instr) is the instrumental resolution, and FWHM(inher) is the inherent line width.

Note.—The uncertainties quoted in this table are dominated by four sources of error: the statistical error due to Poisson noise in the spectrum, the readnoise due to the detector, a systematic error introduced by sky subtraction during the data processing, and the 1σ uncertainties derived for the fit parameters. All of these errors are easily characterized except for the systematic error introduced by sky subtraction. Consequently, we assumed that this additional error is at least as large as the statistical error, and we added it in quadrature to form the total uncertainty.

TABLE 4.2

Diagnostic Emission–Line Ratios

Diagnostic	Flux Ratio		
	HDFX28	CDF–S 202[1]	CXO52[2]
Lyα / C IV λ1549	0.7 ± 0.1	1.66	5.4 ± 0.4
Lyα / Hα(n)	2.4 ± 0.1	$\cdots$	$\cdots$
N V λ1240 / Lyα	1.4 ± 0.1	0.36	0.03:
N V λ1240 / C IV λ1549	0.9 ± 0.1	0.60	0.2:
N V λ1240 / He II λ1640	5.2 ± 0.1	2.11	0.4:
C IV λ1549 / He II λ1640	5.5 ± 0.1	3.54	2.1 ± 0.3
[N II] λ6583 / Hα(n)	1.0 ± 0.1	$\cdots$	$\cdots$

[1]Source: Norman et al. (2002).

[2]Source: Stern et al. (2002b).

TABLE 4.3

PHOTOMETRY OF HDFX28

Observed Bandpass	Rest frame Central λ	Observed Magnitude[†]	Flux Density (μJy)	Detector/ Instrument	Ref.
2–8 keV	18.1 keV	$\cdots$	12×10^{-5}[‡]	*Chandra*/ACIS	1
0.5–2 keV	3.8 keV	$\cdots$	5×10^{-5}[‡]	*Chandra*/ACIS	1
Harris U	1200 Å	24.40	0.33	KPNO 4m/MOSAIC	2
B	1450 Å	24.05	1.02	Keck/LRIS	3
G	1600 Å	24.37	0.69	Palomar 200–inch/COSMIC	4
Kron–Cousins V	1830 Å	23.89	1.01	CFHT/Hawaii 8K CCD Mosaic	5
Kron–Cousins R	2160 Å	23.5	1.2	Keck/LRIS	3
R	2300 Å	23.75	0.87	Palomar 200–inch/COSMIC	4
Kron–Cousins I	2660 Å	22.9	1.8	CFHT/Hawaii 8K CCD Mosaic	3
I_{814}(AB)	2700 Å	23.46	1.43	*HST*/WFPC2	6
HK'	5980 Å	19.3	9.7	UH 2.2m/QUIRC	5
K_s	7140 Å	19.01	17.62	Palomar 200–inch/COSMIC	4
15 μm	5.0 μm	$\cdots$	441^{+43}_{-82}	*ISO*/ISOCAM	7
8.5 GHz	25.6 GHz	$\cdots$	8.15	VLA	8
1.4 GHz	4.2 GHz	$\cdots$	87.8	VLA	9

[†]All magnitudes are normalized to Vega, except the I_{814} magnitude, which is AB. The conversion between Vega–based I–band magnitudes and I_{AB} is $I \approx I_{AB} - 0.3$.

[‡]X–ray flux densities were estimated from the *Chandra*/ACIS hard and soft band fluxes by assuming an X–ray spectral index of $\alpha = +0.7$ (where $F(E) \propto E^\alpha$), based on the hardness ratio described in § 4.3.3.

REFERENCES.—(1) Hornschemeier et al. 2001; (2) C. McNally et al., in preparation; (3) Barger et al. 2000; (4) Hogg et al. 2000; (5) Barger et al. 1999; (6) Mann et al. 1997; (7) Aussel et al. 1999; (8) Richards et al. 1998; (9) Richards 2000.

Part II

Narrow Band Imaging Selection

Chapter 5

Spectroscopic Properties of the z ≈ 4.5 Lyα–emitters

A version of this chapter was previously published in *The Astrophysical Journal* (Dawson, S., Rhoads, J. E., Malhotra, S., Stern, D., Dey, A., Spinrad, H., Jannuzi, B. T., Wang, J., & Landes, E. 2004, ApJ, 617, 707). Reproduced by permission of the AAS.

Abstract

We present Keck/LRIS optical spectra of 17 Lyα–emitting galaxies and one Lyman break galaxy at $z \approx 4.5$ discovered in the Large Area Lyman Alpha (LALA) survey. The survey has identified a sample of ~ 350 candidate Lyα–emitting galaxies at $z \approx 4.5$ in a search volume of 1.5×10^6 comoving Mpc^3. We targeted 25 candidates for spectroscopy; hence, the 18 confirmations presented herein suggest a selection reliability of 72%. The large equivalent widths (median $W_\lambda^{\mathrm{rest}} \approx 80$ Å) but narrow physical widths ($\Delta v < 500$ km s^{-1}) of the Lyα emission lines, along with the lack of accompanying high–ionization state emission lines, suggest that these galaxies are young systems powered by star formation rather than by AGN activity. Theoretical models of galaxy formation in the primordial Universe suggest that a small fraction of Lyα–emitting galaxies at $z \approx 4.5$ may still be nascent, metal–free objects. Indeed, we find with 90% confidence that 3 to 5 of the confirmed sources show $W_\lambda^{\mathrm{rest}} > 240$ Å, exceeding the maximum Lyα equivalent width predicted for normal stellar populations. Nonetheless, we find no evidence for He II $\lambda1640$ emission in either individual or composite spectra, indicating that though these galaxies are young, they are not truly primitive, Population III objects.

5.1 Introduction

Lyα emission has recently begun to fulfill its long–awaited role as a tracer of young galaxies in the high–redshift universe. Although early predictions based on monolithic collapse models (Partridge & Peebles 1967) over–estimated characteristic Lyα line luminosities by factors of ~ 100, the basic insight that Lyα is a good tracer of young stellar populations in association with the gas from which they formed remains valid. Because Lyα photons are resonantly scattered by neutral hydrogen, their effective optical depth to dust absorption in the interstellar medium is greatly enhanced compared to continuum photons of slightly different wavelength. This effect was one of the first proposed to explain non–detections in Lyα protogalaxy searches (Meier & Terlevich 1981), and now that Lyα emission has been detected in high redshift field galaxies (e.g. Hu & McMahon 1996; Cowie & Hu 1998; Dey et al. 1998; Hu et al. 1998; Pascarelle et al. 1998; Hu et al. 1999; Steidel et al. 2000; Kudritzki et al. 2000; Rhoads et al. 2000; Fynbo et al. 2001; Dawson et al. 2001, 2002; Lehnert & Bremer 2003; Bunker et al. 2003; Kodaira et al. 2003), we can turn the effect to our advantage. The line is produced by the interaction of ionizing radiation with hydrogen, and is quenched by dust. Thus, on the simplest interpretation, Lyα–selected samples will likely include galaxies with hot, young stellar populations and little dust, properties which are expected in primitive systems where little chemical evolution has yet occurred.

The Large Area Lyman Alpha (LALA) survey (Rhoads et al. 2000) has recently identified in deep narrowband imaging a large sample of Lyα–emitting galaxies at redshifts $z \approx 4.5$ (Malhotra & Rhoads 2002), $z \approx 5.7$ (Rhoads & Malhotra 2001; Rhoads et al. 2003), and $z \approx 6.5$ (Rhoads et al. 2004). The rest frame equivalent widths ($W_\lambda^{\rm rest}$) of the Lyα emission measured in the narrowband images generally exceed the maximum expected for a normal stellar population, with 60% of the $z \approx 4.5$ sample showing $W_\lambda^{\rm rest} > 240$ Å (Malhotra & Rhoads 2002). Such large equivalent widths suggest that the Lyα–emission is produced in one of two scenarios: (1) young (age $< 10^7$ years) galaxies undergoing star formation in primitive conditions, where metal–free (or low metallicity) gas results in stellar populations biased toward massive, UV–bright stars; or (2) Lyα–emission powered by AGN activity.

We report on the spectroscopic confirmation of 18 narrowband–selected galaxies at $z \approx 4.5$. The narrow physical widths (when resolved, $\Delta v < 500$ km s^{-1}) of the observed Lyα–emission lines rule out conventional broad–lined (Type I) AGN as the central engines of the Lyα–emitters. Moreover, the general lack of accompanying high–ionization state emission lines (e.g. N V $\lambda 1240$, Si IV $\lambda\lambda 1394,1403$, C IV $\lambda 1549$, He II $\lambda 1640$), along with the recent non–detection of the $z \approx 4.5$ sources in deep *Chandra* imaging (Malhotra et al. 2003; Wang et al. 2004), also rules out the comparatively rarer high–redshift narrow–lined (Type II) AGN. These findings leave massive star formation in low metallicity gas as the likely Lyα emission mechanism. The tantalizing limit of such systems — star formation in zero–metallicity gas — represents the first bout of star formation in the pre–galactic Universe, and would be recognizable in optical spectroscopy by weak He II $\lambda 1640$ emission.

We describe our imaging and spectroscopic observations in § 5.2, and we summarize the results of the spectroscopic campaign in § 5.3. In § 5.4, we use our spectroscopic confirmations to update the statistics of the $z \approx 4.5$ population, and we discuss the implications of C IV $\lambda 1549$ and He II $\lambda 1640$ non–detections in composite spectra for the possibility of AGN activity and/or zero–metallicity star formation among the $z \approx 4.5$ sample. Throughout this paper we adopt a Λ–cosmology with $\Omega_M = 0.3$ and $\Omega_\Lambda = 0.7$, and $H_0 = 70$ km s^{-1} Mpc^{-1}. At $z = 4.5$, such a universe is 1.3 Gyr old, the lookback time is 90.2% of the total age of the Universe, and an angular size of $1''.0$ corresponds to 6.61 kpc.

5.2   Observations

5.2.1   Narrowband and Broadband Imaging

The LALA survey concentrates on two primary fields, "Boötes" (14:25:57 +35:32; J2000.0) and "Cetus" (02:05:20 −04:55; J2000.0). Each field is 36×36 arcminutes in size, corresponding to a single field of the 8192×8192 pixel Mosaic CCD camera on the 4m Mayall Telescope at Kitt Peak National Observatory and on the 4m Blanco Telescope at Cerro Tololo Inter–American Observatory. The $z \approx 4.5$ search uses five overlapping narrowband filters each with full width at half maximum (FWHM) ≈ 80 Å. The central wavelengths are 6559, 6611, 6650, 6692, and 6730 Å, giving a total redshift coverage of $4.37 < z < 4.57$ and a survey volume of 7.4×10^5 comoving Mpc^3 per field. In roughly 6 hours per filter per field, we achieve 5σ line detections in $2''.3$ apertures of $\approx 2 \times 10^{-17}$ erg cm^{-2} s^{-1}.

The primary LALA survey fields were chosen to lie within the NOAO Deep Wide Field Survey (NDWFS; Jannuzi & Dey 1999). Thus, deep NDWFS broadband images are available in a custom B_W filter ($\lambda_0 = 4135$ Å, FWHM = 1278 Å) and in the Harris set Kron–Cousins R and I, as well as J, H, K, and K_s. The LALA Boötes field benefits from additional deep V and SDSS z' filter imaging. The imaging data reduction is described in Rhoads et al. (2000), and the $z \approx 4.5$ candidate selection is described in Malhotra & Rhoads (2002). Briefly, candidates are selected based on a 5σ detection in a narrowband filter, the flux density of which must be twice the R–band flux density, and must exceed the R–band flux density at the 4σ confidence level. To guard against foreground interlopers, we set a minimum observed equivalent width of $W_\lambda^{\mathrm{obs}} > 80$ Å, and the candidate must not be detected in the B_W–band.

5.2.2   Spectroscopic Observations

Between 2000 April and 2003 May, we obtained deep spectra of a cross–section of emission line candidates with the Low Resolution Imaging Spectrometer (LRIS; Oke et al. 1995) at the Cassegrain foci of the 10m Keck I and Keck II telescopes (pixel scale $0.215''\mathrm{pix}^{-1}$). The observations were divided between two spectrograph configurations: low resolution, red channel–only observations employing the 150 lines mm^{-1} grating ($\lambda_{\mathrm{blaze}} = 7500$ Å; 4.8 Å pix^{-1} dispersion; $\Delta\lambda_{\mathrm{FWHM}} \approx 25$ Å ≈ 1000 km s^{-1}), and higher resolution observations employing simultaneously the blue–channel 300 lines mm^{-1} grism ($\lambda_{\mathrm{blaze}} = 5000$ Å; 2.64 Å pix^{-1} dispersion; $\Delta\lambda_{\mathrm{FWHM}} \approx 14$ Å ≈ 600 km s^{-1}) and the red–channel 400 lines mm^{-1} grating ($\lambda_{\mathrm{blaze}} = 8500$ Å; 1.86 Å pix^{-1} dispersion; $\Delta\lambda_{\mathrm{FWHM}} \approx 6$ Å ≈ 200 km s^{-1}) with a dichroic splitting the channels at 5000 Å. The data were taken with slitmasks designed to obtain spectra for ~ 15 targets simultaneously; we employed slit widths from $1.2''$ to $1.5''$. Total exposure times ranged from 1.5 hours to 4.3 hours. In each case, the total exposure was broken into a small number of individual integrations between which we performed $\sim 3''$ spatial offsets to facilitate the removal of fringing at long wavelengths. The airmass never exceeded 1.3 during the observations, and the seeing ranged from $0.7''$ to $1.0''$. There was no overlap between the objects targeted for 150ℓ/mm–grating observations and objects targeted for 400ℓ/mm–grating observations.

We used the IRAF[1] package (Tody 1993) to process the data following standard slit

---

[1] IRAF is distributed by the National Optical Astronomy Observatory, which is operated by the Association of Universities for Research in Astronomy, Inc., under cooperative agreement with the National Science Foundation.

spectroscopy procedures. Some aspects of treating the two–dimensional data were facilitated by a custom software package, BOGUS[2], created by D. Stern, A.J. Bunker, and S.A. Stanford. We extracted one–dimensional spectra using the optimal extraction algorithm described in Horne (1986). Wavelength calibrations were performed in the standard fashion using Hg, Ne, Ar, and Kr arc lamps; we employed telluric sky lines to adjust the wavelength zero–point. We performed flux calibrations with longslit observations of standard stars from Massey & Gronwall (1990) taken with the instrument in the same configuration as the corresponding slitmask observation. As the position angle of an observation was set by the desire to maximize the number of targets on a given slitmask, the observations were generally not made at the parallactic angle. Moreover, the data were generally not collected under photometric conditions: five of our six observing runs were affected by light to moderate cirrus.

To investigate the possibility that the sky–subtraction performed during object extraction introduced systematic errors, we made twenty extractions of ostensibly source–free regions in the two–dimensional spectra, employing the same trace used on the neighboring Lyα–emitting candidate. We then fit the resulting "blank–sky" spectra in exactly the same way and over exactly the same wavelength region that we fit for continuum redward and blueward of the emission line in the object extractions (see § 5.3 and Table 5.1). In the complete absence of systematic errors, we expect the blank–sky fits to be zero, barring photon counting statistics. In fact, we find a tiny residual signal: $f_\nu = 0.010$ μJy $\pm$ 0.002 μJy blueward of the emission lines, and $f_\nu = 0.002$ μJy $\pm$ 0.001 μJy redward of the emission lines. These values and their error bars constitute the weighted–averages of the residual signals measured in the 20 extractions, where the weights are the uncertainty in each fit. The error bars represent the uncertainty in the mean residual level (i.e. rather than the scatter among the twenty measurements). The residual signals were subtracted from the object continuum fits, and the errors were adjusted accordingly.

5.3   Spectroscopic Results

Of 25 spectroscopic candidates, 18 were confirmed as galaxies at $z \approx 4.5$. All but one of the 18 confirmed galaxies show Lyα in emission; the remaining galaxy lacks an emission line but shows a large spectral discontinuity identified as the onset of foreground

---

[2]BOGUS is available online at http://zwolfkinder.jpl.nasa.gov/~stern/homepage/bogus.html.

Lyα–forest absorption at $\lambda_{\text{rest}} = 1216$ Å. Of the 7 targets that were not confirmed as Lyα–emitters, 6 were non–detections (to a 5σ upper limit of $\sim 1.1 \times 10^{-17}$ ergs cm^{-2} s^{-1}), and one was a clear [O II] λ3727–emitter at $z = 0.801$, based on the presence of continuum blueward of the emission line. Table 5.1 gives the redshifts of the confirmed Lyα–emitters and summarizes the properties of the emission lines. Figures 5.1 and 5.2 contain the confirming spectra.

We note that each confirmation of a $z \approx 4.5$ Lyα–emitter originates in the spectroscopic detection of a single emission line, the interpretation of which can be problematic. In general, a single, isolated line could be any one of Lyα, [O II] λ3727, [O III] λ5007, or Hα, though given sufficient spectral coverage most erroneous interpretations can be ruled out by the presence of neighboring emission lines: Hβ and [O III] λ4959 for [O III] λ5007; [N II] $\lambda\lambda$6548,6583 and [S II] $\lambda\lambda$6716,6731 for Hα. Hence, the primary threat to determining one–line redshifts is the potential for misidentifying Lyα as [O II] λ3727, or vice versa. Even so, [O II] λ3727 at $z = 0.8$ is generally accompanied by the Hβ plus [O III] $\lambda\lambda$4959,5007 complex, redshifted to ≈ 9000 Å. Such a detection is challenging, owing to heavy contamination by night–sky emission lines in this region of the spectrum. Nonetheless, each spectrum described herein is consistent with an [O III] λ5007 non–detection, with a typical 5σ upper limit to [O III] λ5007 flux of $\sim 3 \times 10^{-17}$ ergs cm^{-2} s^{-1}. One source has a 5σ upper limit of 9×10^{-17} ergs cm^{-2} s^{-1}; the search region for [O III] λ5007 in this case happens to fall directly on prominent sky line residuals. Ignoring this outlier, the scatter in the 5σ upper limits for the remaining 17 sources is just 0.5×10^{-17} ergs cm^{-2} s^{-1}.

5.3.1   Results from the 400ℓ/mm–Grating Observations

The confirmed $z \approx 4.5$ sources observed in our higher–resolution (400ℓ/mm) spectroscopic configuration typically show the asymmetric emission line profile characteristic of high–redshift Lyα, where neutral hydrogen outflowing from an actively star–forming galaxy imposes a sharp blue cutoff and broad red wing (e.g. Stern & Spinrad 1999; Manning et al. 2000; Dawson et al. 2002; Hu et al. 2004; Rhoads et al. 2003). The opposite profile is expected for [O II] λ3727 observed at this resolution (e.g. see Rhoads et al. 2003); hence, the asymmetry typically detected in our higher–resolution sample is good evidence for the Lyα–interpretation.

To quantify this conclusion, we consider two measures of line asymmetry (see Rhoads

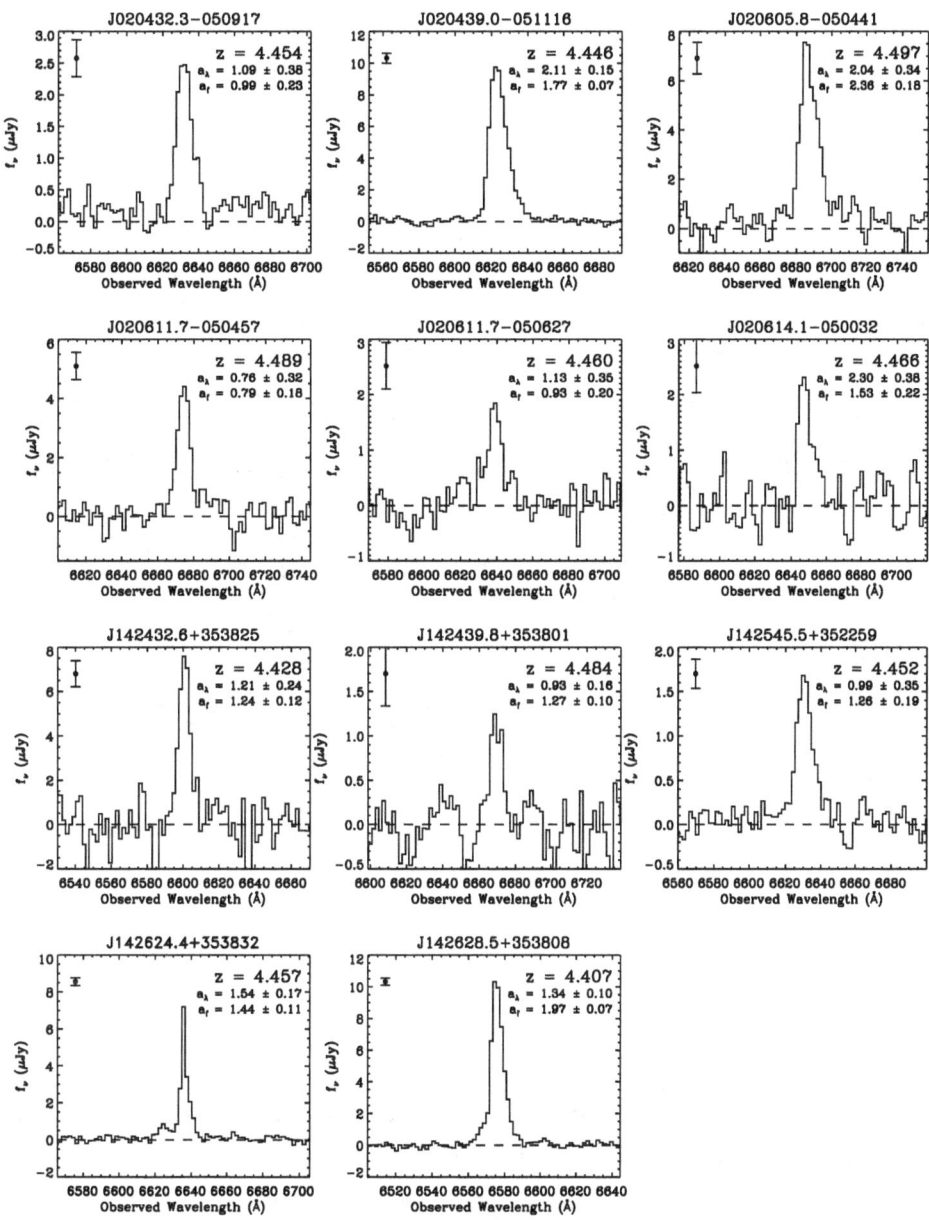

FIG. 5.1.— Spectra of the 11 confirmed $z \approx 4.5$ galaxies observed with the Keck/LRIS 400ℓ/mm-grating, with a wavelength range selected to highlight the emission line profile. The measured redshifts and asymmetry statistics (§ 5.3) are indicated at the upper right. The representative error bar (upper left) is the median of the flux errors in each pixel over the wavelength range displayed. The spectra are unsmoothed.

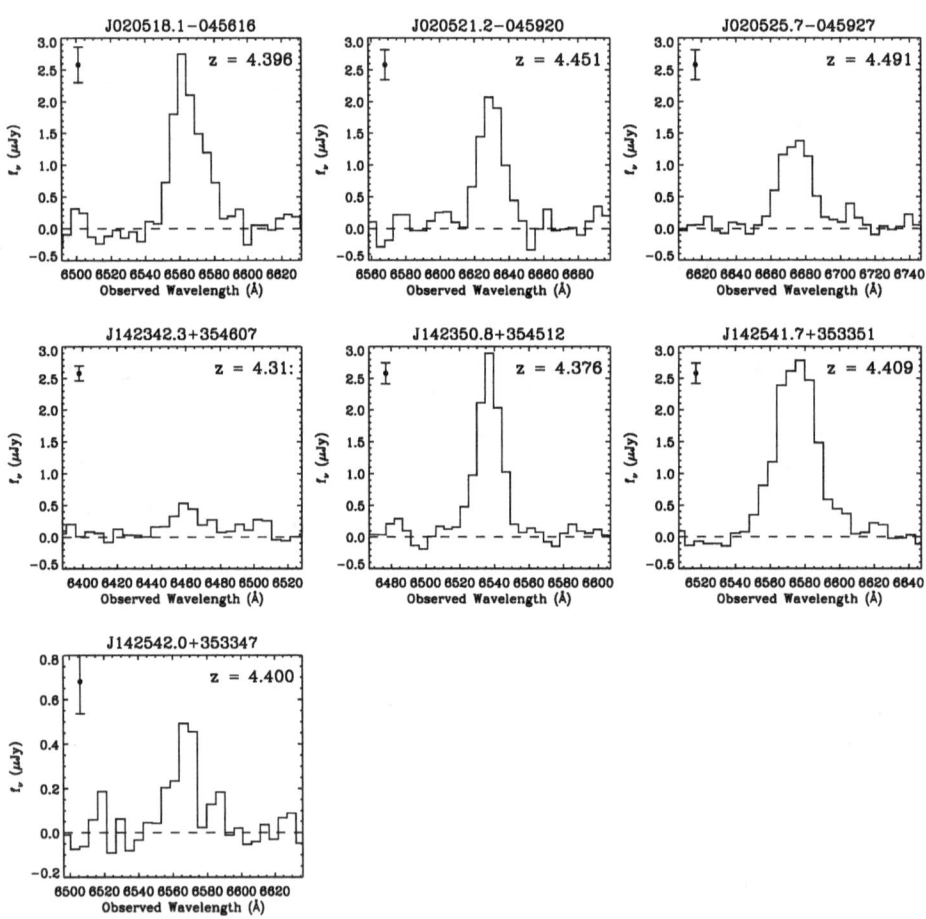

Fig. 5.2.— Spectra of the 7 confirmed $z \approx 4.5$ galaxies observed with the Keck/LRIS 150ℓ/mm–grating, with a wavelength range selected to highlight the emission line profile. The measured redshifts are indicated at the upper right. The representative error bar (upper left) is the median of the flux errors in each pixel over the wavelength range displayed. The spectra are unsmoothed.

et al. 2003, 2004). For both, we determine the wavelength of the peak of the emission line (λ_p), and where the line flux exceeds 10% of the peak on the blue side ($\lambda_{10,b}$) and on the red side ($\lambda_{10,r}$). The "wavelength ratio" is then $a_\lambda = (\lambda_{10,r} - \lambda_p)/(\lambda_p - \lambda_{10,b})$, and the "flux ratio" is $a_f = \int_{\lambda_p}^{\lambda_{10,r}} f_\lambda d\lambda / \int_{\lambda_p}^{\lambda_{10,b}} f_\lambda d\lambda$[3]. As in Rhoads et al. (2004), we experimented with raising and lowering the flux threshold but found no clear benefit to using values other than 10%. Lowering the threshold results in enhanced contamination from continuum noise, increasing the scatter and uncertainty in the measurements; raising the threshold diminishes the contribution of the broad, red Lyα wing to the measurement, reducing the ability of the asymmetry measures to discriminate Lyα from symmetric interlopers. Figure 5.3 compares the distribution in a_f–a_λ space of 28 $z \sim 1$ [O II] λ3727–emitters (provided by the DEEP2 team; Davis et al. 2003, A. Coil 2004, private communication) to 22 confirmed and putative high–redshift Lyα–emitters. As a population, the Lyα–emitters are systematically segregated from the [O II] λ3727–emitters. While the [O II] λ3727–emitters are distributed according to $a_f = 0.8 \pm 0.1$ and $a_\lambda = 0.8 \pm 0.1$, all save one of the Lyα–emitters satisfies $a_f > 1.0$ or $a_\lambda > 1.0$, and more than half satisfy both.

We note that one source included herein as a confirmed Lyα–emitter (J020611.7−050457) falls in the region of a_f–a_λ space characteristic of the [O II] λ3727–emitters. With $a_f = 0.79 \pm 0.10$ (1σ) and $a_\lambda = 0.76 \pm 0.05$ (1σ), the asymmetry measures do not favor the Lyα–interpretation for this source, while the lack of blue continuum emission, on the other hand, does not favor the [O II] λ3727–interpretation. The large equivalent width measured for the line is similarly ambiguous. For Lyα, the redshift is $z = 4.489$ and the *observed* frame equivalent width of $W_\lambda^{\rm obs} = 620$ Å $\pm$ 200 Å (1σ) implies a rest frame equivalent width of $W_\lambda^{\rm rest} = 111$ Å $\pm$ 37 Å (1σ). This value is consistent with $W_\lambda^{\rm rest}$ for other high–redshift Lyα–lines presented here and elsewhere (Hu & McMahon 1996; Cowie & Hu 1998; Hu et al. 1998; Malhotra & Rhoads 2002; Rhoads et al. 2003). Alternatively, interpreting the emission line as [O II] λ3727 implies a rest frame equivalent width of $W_\lambda^{\rm rest} = 340$ Å. Such a large value for [O II] λ3727 is rare but not wholly unprecedented: large continuum–selected (Cowie et al. 1996; Hogg et al. 1998) and Hα–selected (Gallego et al. 1996) samples indicate that [O II] λ3727 very rarely exceeds $W_\lambda^{\rm rest} > 100$ Å, but at least one exceptional source with $W_\lambda^{\rm rest} \approx 600$ Å at $z = 1.464$ is known (Stern et al. 2000b). The [O II] λ3727–

---

[3]The error bars on a_λ and a_f were determined with Monte Carlo simulations in which we modeled each emission line with the truncated Gaussian profile described in Hu et al. (2004) and Rhoads et al. (2004), added random noise in each pixel according to the photon counting errors, and measured the widths $\sigma(a_\lambda)$ and $\sigma(a_f)$ of the resulting distributions of a_λ and a_f for the given line. That is, for each $a_{\lambda,i}$, the error $\delta a_{\lambda,i} = \sigma(a_{\lambda,i})$, and similarly for each $a_{f,i}$.

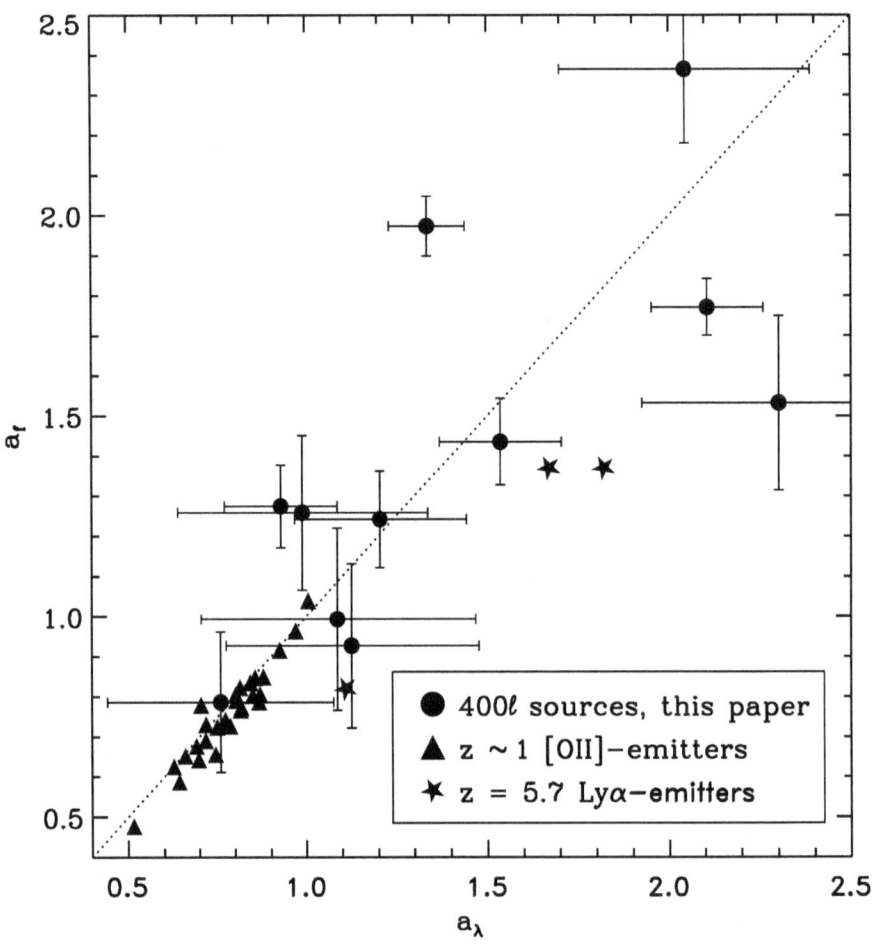

FIG. 5.3.— Scatter plot of the flux–based asymmetry statistic a_f vs. the wavelength–based asymmetry statistic a_λ for known high–redshift Lyα–emitters, and for a sample of [O II] λ3727–emitters at $z \sim 1$. The 11 Lyα–emitters at $z \sim 4.5$ are the 400ℓ/mm–grating observations presented in this paper. (The 150ℓ/mm–grating observations, being of much lower resolution, are not included.) The 3 Lyα–emitters at $z = 5.7$ are from Rhoads et al. (2003). The 28 [O II] λ3727–emitters at $z \sim 1$ were provided by the DEEP2 team (Davis et al. 2003, A. Coil 2004, private communication); their Keck/DEIMOS 1200ℓ/mm–grating spectra were smoothed to the Keck/LRIS 400ℓ/mm–grating resolution by convolution with a Gaussian kernel.

interpretation would further imply the presence of [O III] $\lambda5007$ at a redshifted wavelength of 8980 Å; no such line is detected to a 5σ upper limit of $< 2.3 \times 10^{-17}$ ergs cm^{-2} s^{-1}. Though we tentatively include this source among our confirmed Lyα–emitters, the reader is encouraged to keep the foregoing caveat concerning its interpretation in mind.

5.3.2   Results from the 150ℓ/mm–Grating Observations

Our lower resolution (150ℓ/mm) spectroscopic configuration trades sensitivity to line shape for sensitivity to continuum. Therefore, while the (typically unresolved) line profiles in this sample are less useful for discriminating Lyα from [O II] $\lambda3727$, the continuum detection (if any) can be used to measure the amplitude of the discontinuity at the emission line. On the Lyα–interpretation, we expect a continuum break owing to the onset of absorption by the Lyα–forest at $\lambda_{\rm rest} = 1216$ Å, a robust spectral signature used extensively in the photometric selection of galaxies at $z > 5$ (e.g. Dey et al. 1998; Weymann et al. 1998; Spinrad et al. 1998; Lehnert & Bremer 2003). The likely low–redshift interloper in this case is the 4000 Å break [$D(4000)$], resulting from the sudden onset of stellar photospheric opacity shortward of 4000 Å. We characterize the break amplitude in the 150ℓ/mm–grating observations with $1 - f_\nu^{\rm short}/f_\nu^{\rm long}$, where $f_\nu^{\rm short}$ is the variance–weighted average flux density in a 1200 Å window beginning 30 Å below the emission line, and $f_\nu^{\rm long}$ is the same, but above the emission line. The median of the 1σ *lower* limits to the amplitude of the flux decrement in our $z \approx 4.5$ candidates, calculated from Monte Carlo simulations of the flux densities with the constraint $f_\nu > 0$, is then $1 - f_\nu^{\rm short}/f_\nu^{\rm long} > 0.61$. This value is entirely consistent with both theoretical models of the Lyα break amplitude at $z = 4.5$ (e.g. Madau 1995; Zhang et al. 1997) and with measurements from existing data sets (see Stern & Spinrad 1999, and references therein). By contrast, a sample of 43 galaxies in the redshift range $0.7 < z < 0.94$ (roughly corresponding to the resulting redshift if [O II] $\lambda3727$ has been misidentified as Lyα) has $D(4000)$ amplitudes of $1 - f_\nu^{\rm short}/f_\nu^{\rm long} = 0.39 \pm 0.1$ (Dressler & Gunn 1990). We take Lyα as the most likely line identification under these circumstances.

We present an unweighted coaddition of the 11 400ℓ/mm spectra in Figure 5.4 and of the 7 150ℓ/mm spectra in Figure 5.5. With its higher resolution, the 400ℓ/mm composite spectrum highlights the asymmetry of the Lyα emission line. We find $a_f = 1.58\pm0.09\,(1\sigma)$ and $a_\lambda = 1.69 \pm 0.21\,(1\sigma)$ for the composite. With its greater sensitivity to continuum, the 150ℓ/mm composite highlights the spectral discontinuity at the emission line. We find $1 - f_\nu^{\rm short}/f_\nu^{\rm long} = 0.75 \pm 0.06\,(1\sigma)$ for the composite. The rest frame equivalent

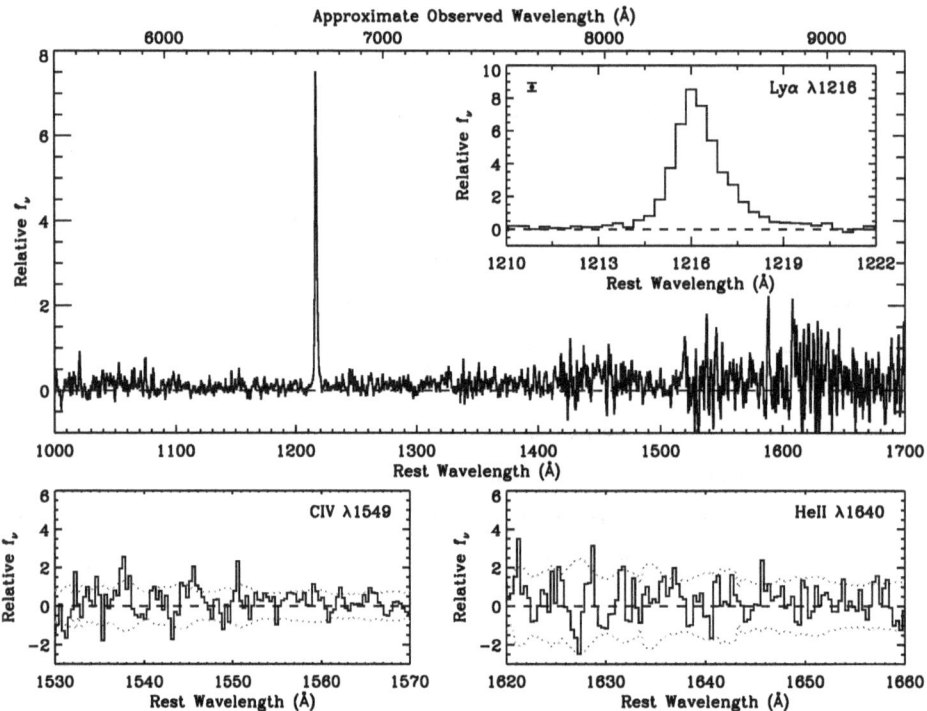

FIG. 5.4.— Composite spectrum consisting of an unweighted coaddition of the 11 $z \approx$ 4.5 galaxies confirmed in 400ℓ/mm–grating observations. The full spectrum (top) has been smoothed with a 3 pixel boxcar filter; the inset, highlighting the asymmetry in the composite Lyα profile, is unsmoothed. The representative error bar (upper left in inset) is the median of the flux errors in each pixel over the wavelength range displayed. The small plots at bottom demonstrate the absence of any significant emission from C 4 λ1549 or He 2 λ1640 in the composite. To wit, the C 4 λ1549 flux is constrained to be < 8% (12%) of the flux in Lyα to a confidence of 2σ (3σ); the He 2 λ1640 flux is constrained to be < 13% (20%) of the flux in Lyα to a confidence of 2σ (3σ). The dotted lines indicate the photon counting errors as they were propagated through the coaddition.

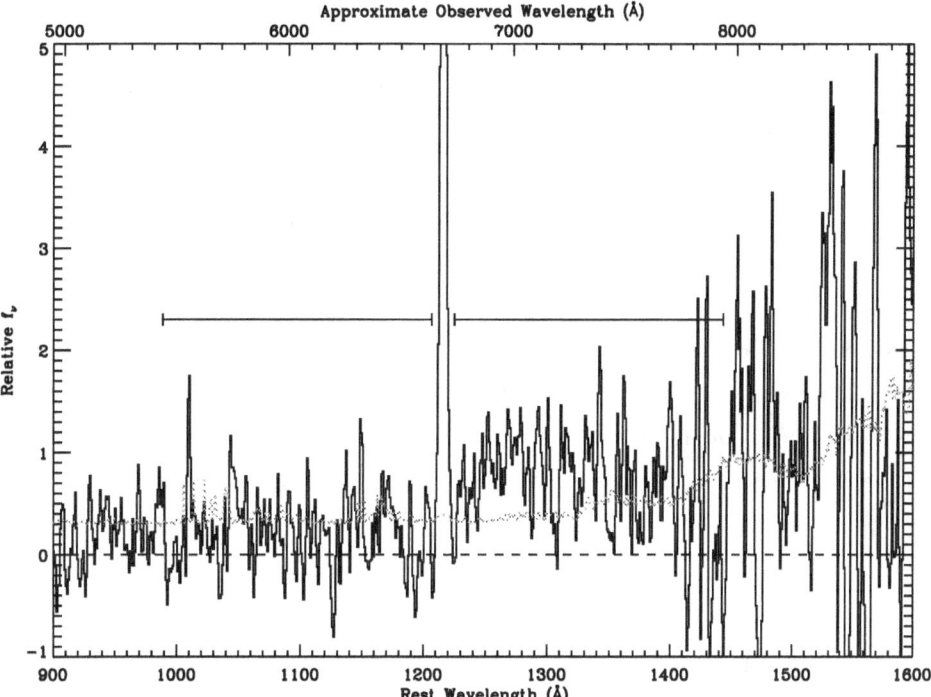

FIG. 5.5.— Composite spectrum consisting of an unweighted coaddition of the 7 $z \approx 4.5$ galaxies confirmed in $150\ell/\mathrm{mm}$–grating observations, with the ordinate scale selected to emphasize the continuum break across the emission line. The spectrum has been smoothed with a 3 pixel boxcar filter. The dotted line indicates the photon counting errors as they were propagated through the coaddition, and accounts for the smoothing. The horizontal bars demarcate the wavelength region considered for the determination of $1 - f_\nu^{\mathrm{short}}/f_\nu^{\mathrm{long}}$, which is equal to 0.75 ± 0.06 (1σ) for the composite.

widths of the composite emission lines are $W_\lambda^{\rm rest}(400\ell/{\rm mm}) = 100$ Å $\pm$ 13 Å (1σ) and $W_\lambda^{\rm rest}(150\ell/{\rm mm}) = 74$ Å $\pm$ 13 Å (1σ), respectively.

5.3.3   Spectroscopic Non–Detections

Of the six spectroscopic non–detections, three candidates fell on slitmasks observed under adverse conditions for which the general spectroscopic yield was low. As such, our failure to confirm these targets as $z \approx 4.5$ Lyα–emitters should not be interpreted as a reflection on the efficacy of candidate selection. The remaining three non–detections were observed under acceptable or photometric conditions, for which the general spectroscopic yield was high. However, each of these targets was suboptimal for one of a variety of reasons: one target sits on a weak satellite residual; one target is very close to a bright star; one target appears in an initial epoch of imaging but not in subsequent epochs and is therefore likely a variable source or a spurious detection. For the remainder of this paper, we will cite a selection reliability of 72%, but the reader is cautioned that this is the most conservative estimation; it does not discriminate between spurious candidates produced at the stage of narrowband–selection, or failures in the spectroscopic follow–up. That is, this figure for the selection reliability does not imply that 28% of candidate Lyα–emitters in narrowband imaging surveys are unsound.

5.4   Discussion

5.4.1   The Statistics of the $z = 4.5$ Population

The spectroscopic confirmation of 17 $z \approx 4.5$ Lyα–emitters out of 25 candidates allows us to update the statistics of this population as they were presented in Malhotra & Rhoads (2002). By applying a 0.72 correction factor to the observed source counts, we find a number density of ≈ 2800 Lyα–emitters per square degree per unit redshift above a detection threshold of 2×10^{-17} ergs cm^2 s^{-1}. This figure translates to a Lyα–luminosity density at $z \approx 4.5$ of $2 \times 10^5 L_\odot$ Mpc^{-3} for sources with $L_{\rm Ly\alpha} > 1.04 \times 10^9 L_\odot$. Hu et al. (1999) give a conversion factor of 1 $M_\odot$ yr$^{-1} = 10^{42}$ ergs s^{-1} for converting Lyα–luminosities into star formation rates (but see the caveats in Rhoads et al. 2003); together with the Lyα–luminosity density estimate, this yields a star formation rate density at $z \approx 4.5$ of $8 \times 10^{-4} M_\odot$ yr^{-1} Mpc^{-3}, with individual star formation rates ranging from 1 $M_\odot$ yr^{-1} to 16 $M_\odot$ yr^{-1}. These results are roughly consistent with previous studies with similar limiting fluxes (e.g. Hu et al. 1998; Ouchi et al. 2003).

5.4.2   The Equivalent Width Distribution

The rest frame equivalent widths reported for each source in Table 5.1 were determined directly from the spectra according to $W_\lambda^{\rm rest} = (F_\ell/f_{\lambda,r})/(1+z)$, where F_ℓ is the flux in the emission line and $f_{\lambda,r}$ is the measured red–side continuum flux density. For the one case with $f_{\lambda,r}$ formally consistent with zero, we derive a 2σ lower limit to the equivalent width. The resulting equivalent width distribution is plotted in Figure 5.6, with individual measurements keyed to the grating with which each source was observed. There is no obvious systematic difference between equivalent widths measured in our lower resolution spectroscopic configuration and those measured in our higher resolution configuration.

Using Starburst99 models (Leitherer et al. 1999) with a Salpeter initial mass function (IMF), an upper mass cutoff of 120 $M_\odot$, and a metallicity of 1/20th solar, Malhotra & Rhoads (2002) calculated maximum equivalent widths of 300 Å for a one Myr–old stellar population, 150 Å for a 10 Myr–old population, and 100 Å for populations older than 10^8 years. Owing to the lower metallicity, these values are slightly higher than the previous calculations of Charlot & Fall (1993), who present 240 Å as the highest equivalent width achievable by a stellar population.

To properly compare these model predictions to the equivalent width distribution presented herein, one must first account for absorption in the intergalactic medium (IGM), which affects the observations but is not included in the models. Equivalent widths determined from spectroscopy are based on the flux in the observed emission line (which may be affected by intergalactic absorption), and on the red–side continuum flux density (which is unaffected by intergalactic absorption). Malhotra & Rhoads (2002) used the prescription for IGM absorption given by Madau (1995) to derive a flux decrement of a factor of 0.64 for an intrinsically symmetric Lyα emission line centered on zero velocity. Of course, this calculation neglects the effects of the dust content and detailed kinematics of the galaxy ISM, which increasingly appear to play a significant role in determining the emergent Lyα profile (e.g. Kunth et al. 1998; Stern & Spinrad 1999; Mas–Hesse et al. 2003; Shapley et al. 2003; Ahn 2004). Nonetheless, adopting this correction factor as an upper limit on possible IGM effects dictates that the limiting model equivalent widths measured spectroscopically would be 190 Å, 100 Å, and 60 Å for populations of ages $10^6, 10^7$ and 10^8 years respectively.

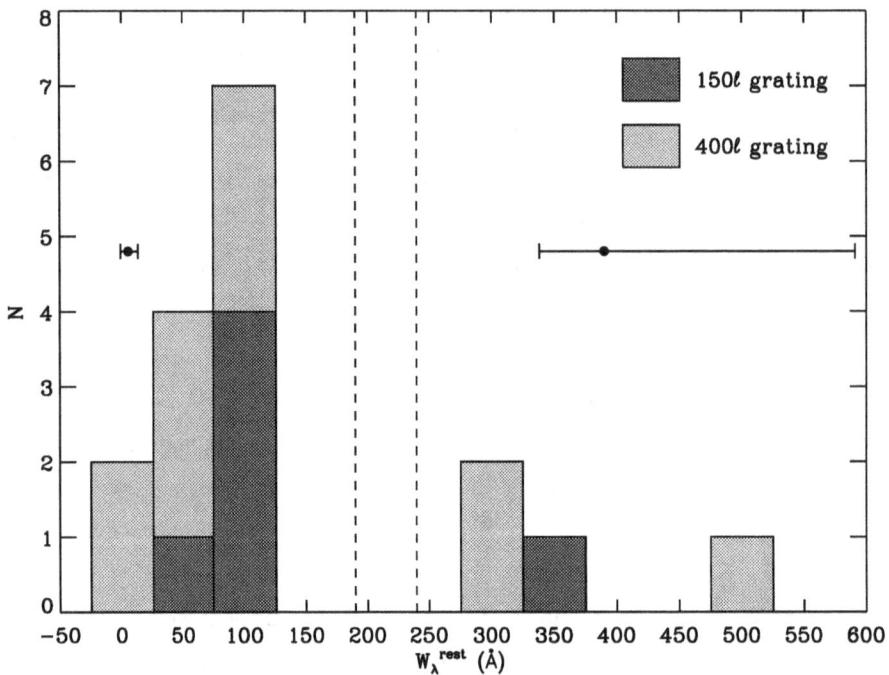

Fig. 5.6.— Histogram of the spectroscopic rest frame equivalent widths for the $z \approx 4.5$ population, determined with $W_\lambda^{rest} = (F_\ell/f_{\lambda,r})/(1+z)$, where F_ℓ is the flux in the emission line and $f_{\lambda,r}$ is the measured red–side continuum flux density. The dashed vertical lines mark the maximum Lyα equivalent widths predicted by the stellar models of Charlot & Fall (1993) and Malhotra & Rhoads (2002). Representative error bars on the equivalent widths are plotted at left and at right. Notably, the highest equivalent widths are generally the least certain, as they correspond to the faintest (and hence least certain) continuum estimates.

A second concern regarding the equivalent width determination in spectroscopy is the high sensitivity to uncertainty in the continuum measurement. Since the continuum estimate enters into the denominator of the expression for $W_\lambda^{\mathrm{rest}}$, the characteristically small continuum values, along with their considerable error bars, tend to cause large scatter in the results. Not surprisingly, the sources in our sample with the largest measured equivalent widths tend also to have the largest fractional uncertainty therein.

With these caveats in mind, we performed a careful statistical analysis designed to place rigorous constraints on the number of galaxies in our sample with equivalent widths in excess of the maximum values allowed by stellar population models. First, we associated each measured line flux $F_{\ell,i} \pm \delta F_{\ell,i}$ with a Gaussian probability density function (PDF) centered on $F_{\ell,i}$ with width $\sigma = \delta F_{\ell,i}$; we proceeded similarly for the measured continuum fluxes. We then generated a grid of line flux versus continuum flux on which each node has an associated equivalent width and is assigned a weight according to the Gaussian error distribution on each of its fluxes. Next we collapsed the grid into a histogram of equivalent widths, adding the weight from each grid point to the appropriate equivalent width bin. The result is a non–Gaussian PDF $P_i(w)$ for which $P_i(w)\,dw$ is the probability of observing $W_{\lambda,i}^{\mathrm{rest}}$ in the interval $w < W_{\lambda,i}^{\mathrm{rest}} < w+dw$. We used the final ensemble of $P_i(w)$ to determine the likelihood that exactly N galaxies in our sample exceed some limiting $W_\lambda^{\mathrm{rest}}$, and we added these likelihoods to determine the confidence with which a range of N galaxies exceed that $W_\lambda^{\mathrm{rest}}$ limit.

Proceeding in this fashion, we find with 90% confidence that 3 to 5 galaxies in our sample exceed the fiducial Charlot & Fall (1993) upper limit of $W_\lambda^{\mathrm{rest}} > 240$ Å, and with 91% confidence that 4 to 6 galaxies exceed $W_\lambda^{\mathrm{rest}} > 190$ Å, the upper limit of Malhotra & Rhoads (2002), with the largest reasonable correction for IGM absorption. Thus, no matter how one treats the effects of IGM absorption on the maximum equivalent widths predicted by stellar population models, we find a significant fraction of our sample in excess of those limits. These galaxies are required to be very young (age $< 10^6$ years), or to have skewed IMFs, or perhaps to harbor AGN producing stronger–than–expected Lyα emission.

5.4.3   AGN Among the z ≈ 4.5 Population?

Given the large equivalent widths measured for the high–redshift Lyα emission in this sample and elsewhere (Kudritzki et al. 2000; Rhoads et al. 2003, 2004), one intriguing scenario is the possibility that our narrowband selection has identified a large population of high–redshift AGN. However, the recent non–detections in deep (~ 170 ks) *Chandra*/ACIS imaging of the narrowband–selected sources in the LALA Boötes field (Malhotra et al. 2003) and the LALA Cetus field (Wang et al. 2004) have placed strong constraints against AGN activity among the Lyα–emitters. No individual Lyα candidate was detected with *Chandra*/ACIS to a 3σ limiting X–ray luminosity of $L_{2-8\mathrm{keV}} = 3.3 \times 10^{43}$ erg s^{-1}. The constraint is even stronger in the stacked X–ray image, which suggests a 3σ limit to the average X–ray luminosity of $L_{2-8\mathrm{keV}} < 2.8 \times 10^{42}$ erg s^{-1}. This limit is roughly an order of magnitude fainter than what is typically observed for even the heavily obscured, Type II AGN (e.g. Stern et al. 2002b; Norman et al. 2002; Dawson et al. 2003).

The case against AGN activity among the Lyα–emitters is further borne out by the optical spectroscopy presented herein. The narrow physical widths of the Lyα emission lines ($\Delta v < 500$ km s^{-1}) definitively rule out conventional broad–lined (Type I) AGN, and also disfavor narrow–lined (Type II) AGN, which have typical $\Delta v_{\mathrm{Ly}\alpha} \sim 1000$ km s^{-1}. Furthermore, no individual spectrum shows evidence of the high–ionization state UV emission lines symptomatic of AGN activity (e.g. N V λ1240, C IV λ1549, He II λ1640) nor is there evidence of such lines in the composite spectra (Figures 5.4 and 5.5). In particular, C IV λ1549 flux in the 400ℓ/mm composite spectrum is constrained to be $\lesssim 8\%$ (2σ) of the flux in Lyα, implying a flux ratio of $f(\mathrm{Ly}\alpha)/f(\mathrm{C\ IV}\ λ1549) \gtrsim 13$. By contrast, the three Type II AGNs cited above span only $0.7 < f(\mathrm{Ly}\alpha)/f(\mathrm{C\ IV}\ λ1549) < 5.4$.

5.4.4   Population III Among the z ≈ 4.5 Population?

The identification of a population of large equivalent width Lyα–emitters evidently powered by star formation in low metallicity gas suggests that we are closing the gap between the first, little–enriched primordial galaxies and the higher metallicities of massive galaxies in the local universe. Indeed, recent numerical studies of the rest frame UV and optical properties of very low metallicity stellar populations indicate that Lyα emission increases strongly with decreasing metallicity, far exceeding $W_\lambda^{\mathrm{rest}} \gtrsim 500$ Å for $Z < 10^{-5} Z_\odot$ (e.g. Schaerer 2003). The tantalizing limit of such studies is star formation in zero metallicity gas, the so–called Population III, which constitutes the first bout of star formation in the pre–galactic Universe.

The striking features of massive Population III stars are their high effective temperatures ($T_{\mathrm{eff}} \sim 10^5$ K for $M > 100 M_\odot$) and consequent hard ionizing spectra, resulting in the production of 60% more H 1–ionizing photons than their Population II counterparts, and up to 10^5 times more He 2–ionizing photons (Tumlinson et al. 2003). As a consequence, in addition to high equivalent width Lyα emission, a unique observational signature of this primeval population is emission from He II recombination lines. These lines are particularly attractive for a detection experiment since they suffer minimal effects of scattering by gas and benefit from decreasing attenuation by dust. That said, the possibility for the direct detection of He II λ1640 in the present data set boils down to two questions: Is Population III star–formation occurring at the comparatively late epoch occupied by the $z \approx 4.5$ Lyα–emitters? And if so, is the resulting He 2 emission of sufficient luminosity to be detected?

As for the first question, if star formation feedback in the Universe is sufficiently weak that metal production proceeds very inhomogeneously, then Population III star formation may continue to surprisingly low redshifts, occurring in regions that have not yet been polluted by previous episodes of star formation. To this end, the analytical models of Scannapieco et al. (2003) designed to investigate the detectability of primordial star formation indicate that Population III objects tend to form in the $10^{6.5}$–$10^{7} M_\odot$ mass range, just large enough to cool within a Hubble time, but small enough that they are not clustered near areas of previous star formation. The result is that somewhere between 1% and 30% of strong Lyα–emitters at $z = 4.5$ should owe their Lyα emission to Population III star formation, depending on the strength of feedback in these systems.

As for the detectability of the resulting He II emission, early predictions of Lyα and He II recombination luminosities in metal–free stellar clusters suggest He II λ1640 emission in excess of 13% that of Lyα, with $W_\lambda^{\mathrm{rest}}(\mathrm{He\ II}\ \lambda1640) \sim 1100$ Å (Bromm et al. 2001). Notably, Schaerer (2002) points out that the prediction of such a large equivalent width likely results from neglecting the contribution of nebular continuum emission to the model Population III spectrum, which dominates the SED for $\lambda > 1400$ Å and therefore reduces the expected equivalent width. More recent models are indeed more conservative in their predictions of the strength of He II emission. Tumlinson et al. (2003) suggest He II fluxes for a Population III cluster at $z \approx 4.5$ ranging from $\sim 0.001\%$ to 5% of the flux in Lyα, where the low extremum is set by the limiting case of an instantaneous starburst, and the high extremum is set by constant star formation at rate of $\sim 40\ M_\odot\ \mathrm{yr}^{-1}$. Schaerer (2003) predicts He II λ1640 equivalent widths in excess of 80 Å for very young zero–metallicity instantaneous starbursts, though at ages greater than ~ 1 Myr, values $W_\lambda^{\mathrm{rest}}(\mathrm{He\ II}\ \lambda1640) \gtrsim 5$ Å are expected only at metallicities $Z < 10^{-7} Z_\odot$.

Physically, the best prospect for detecting He II $\lambda1640$ in our spectroscopic sample lies with the spectra containing Lyα at the highest equivalent widths. However, He II $\lambda1640$ at the suggested flux levels is undetectable in any of our individual spectra, leaving only the possibility that a constraint to He II $\lambda1640$ emission may be derived from the composite spectrum, where we benefit from a $\sqrt{N}$–reduction in Poisson noise. Accordingly, we performed Monte Carlo simulations aimed at measuring He II $\lambda1640$ emission in the $400\ell/\mathrm{mm}$ composite, searching over a distribution of potential He II $\lambda1640$ line widths set by the width of the composite Lyα line. The result is a He II $\lambda1640$ flux which is formally consistent with zero, with a 2σ (3σ) upper limit of 13% (20%) of the flux in Lyα. The corresponding 2σ (3σ) upper limit to the He II $\lambda1640$ rest frame equivalent width is 17 Å (25 Å). This equivalent width limit is sufficient to rule out the youngest zero–metallicity instantaneous burst and continuous SFR models of Schaerer (2003), though metallicities of $Z < 10^{-7}Z_\odot$ are still permissible. We therefore conclude that this data set cannot corroborate the proposition (e.g. Tumlinson et al. 2003) that the high equivalent widths of the $z \approx 4.5$ narrowband–selected Lyα–emitters suggest that the first metal–free stars have already been found.

5.5   Conclusion

Out of 25 narrowband–selected candidates, we have spectroscopically confirmed 18 galaxies at $z \approx 4.5$, implying a selection reliability of 72%. The resulting sample of confirmed Lyα emission lines show large equivalent widths (median $W_\lambda^{\mathrm{rest}} \approx 80$ Å) but narrow physical widths ($\Delta v < 500$ km s^{-1}), supporting the conclusion of Malhotra et al. (2003) and Wang et al. (2004) that the Lyα emission in these sources derives from star formation, not from AGN activity. Moreover, though the expectation from theoretical models of galaxy formation in the primordial Universe is that a small fraction of Lyα–emitting galaxies at $z \approx 4.5$ may be nascent, metal–free objects, we do not detect He II $\lambda1640$ emission in either individual or composite spectra, indicating that though these galaxies are young, they show no evidence of being truly primitive, Population III objects. Of course, this last result may be a function of our comparatively small sample size, which could only be reasonably expected to yield a He II $\lambda1640$ detection if the frequency of Population III objects among $z \approx 4.5$ Lyα–emitters exceeds $\sim 6\%$. Clearly, increasing our sample size with future spectroscopy will provide a far tighter constraint on the make–up of the

$z \approx 4.5$ galaxy population.

One heretofore unexplored consideration is the possibility that galactic–scale winds may be required to work in concert with sub–solar metallicities to facilitate the escape of Lyα radiation from a system (e.g. Kunth et al. 1998; Ahn 2004). In this scenario, the low metallicity acts to reduce the dust opacity, and the wind acts to Doppler shift the absorbers, minimizing resonant scattering of Lyα photons. The detailed geometry of the interstellar medium (ISM) doubtless plays several roles in this process. As one example, a galactic wind driven by star formation has its origin in an over–pressured cavity of hot gas inside the star–forming galaxy. This superbubble ultimately expands and bursts out into the galaxy halo; naturally this expansion, and hence the galactic wind, proceeds in the direction of the vertical pressure gradient (Heckman et al. 2000; Tenorio–Tagle et al. 1999). As a second example, if dust and neutral gas in the ISM have a high covering factor but a low volume filling factor, it is possible for continuum radiation to be strongly absorbed while an appreciable fraction of Lyα line radiation escapes (Neufeld 1991).

Observationally, the extent to which the emission of Lyα photons in a star forming galaxy depends on not only the distribution and kinematics of gas and dust in its ISM, but also on the inclination of the system, remains an open question. Since significant Lyα emission in SCUBA sources (Chapman et al. 2003) offers evidence that the deleterious effect of dust on Lyα emission may be mitigated by strong starburst–driven winds, it appears unlikely that spectroscopy of the rest frame UV of high redshift Lyα–emitters alone will be able to fully disentangle these effects. By contrast, a detailed understanding of the rest frame optical properties of the these systems would offer a strong lever arm on their dust contents and star formation histories. Hence, resolution of these issues may await infrared observations of the $z \approx 4.5$ galaxy sample.

Acknowledgements

This work benefited greatly from conversations with T. Robishaw, J. Simon, D. Schaerer, E. Green, and the anonymous referee. We further acknowledge J. G. Cohen and C. C. Steidel for supporting LRIS–R and LRIS–B, respectively, and we thank the DEEP2 team for providing the sample [O II] λ3727 spectra. In addition, we wish to acknowledge the significant cultural role that the summit of Mauna Kea plays within the indigenous Hawaiian community; we are fortunate to have the opportunity to conduct observations from this mountain. The work of SD was supported by IGPP–LLNL University Collaborative Research Program grant #03–AP–015, and was performed under the auspices of the U.S. Department of Energy, National Nuclear Security Administration by the University of California, Lawrence Livermore National Laboratory under contract No. W–7405–Eng–48. The work of DS was carried out at the Jet Propulsion Laboratory, California Institute of Technology, under contract with NASA. AD and BJ acknowledge support from NOAO, which is operated by the Association of Universities for Research in Astronomy, Inc. under cooperative agreement with the National Science Foundation (NSF). HS gratefully acknowledges NSF grant AST 95–28536 for supporting much of the research presented herein. This work made use of NASA's Astrophysics Data System Abstract Service.

TABLE 5.1

SPECTROSCOPIC PROPERTIES OF THE KECK/LRIS $z \approx 4.5$ LYα-EMITTERS

Target	z[a]	Lyα Flux[b]	$W_\lambda^{\mathrm{rest}}$[c] (Å)	FWHM[d] (Å)	Δv[e] (km s^{-1})	Cont. (μJy)[f] Blue Side	Cont. (μJy)[f] Red Side
Data obtained with the 150ℓ/mm–grating:							
J020518.1−045616	4.396	3.79 ± 0.34	80^{+17}_{-13}	18.1 ± 1.1	< 1140[g]	0.028 ± 0.018	0.126 ± 0.022
J020521.2−045920	4.451	2.52 ± 0.22	79^{+24}_{-15}	16.8 ± 1.1	< 1130[g]	0.030 ± 0.016	0.086 ± 0.019
J020525.7−045927	4.491	2.15 ± 0.24	55^{+13}_{-10}	22.0 ± 1.1	< 1120[g]	0.017 ± 0.016	0.106 ± 0.020
J142342.3+354607	4.31:	⋯	⋯	⋯	⋯	0.032 ± 0.008	0.152 ± 0.009
J142350.8+354512	4.376	3.40 ± 0.18	82^{+94}_{-30}	15.2 ± 1.1	< 1150[g]	0.021 ± 0.011	0.110 ± 0.085
J142541.7+353351	4.409	5.83 ± 0.25	335^{+171}_{-125}	26.3 ± 1.1	< 370	0.041 ± 0.010	0.046 ± 0.034
J142542.0+353347	4.400	0.66 ± 0.16	111^{+201}_{-47}	12.9 ± 1.1	< 1140[g]	0.009 ± 0.009	0.016 ± 0.012
Data obtained with the 400ℓ/mm–grating:							
J020432.3−050917	4.454	1.96 ± 0.14	29^{+3}_{-2}	10.2 ± 0.3	< 370	0.069 ± 0.015	0.182 ± 0.015
J020439.0−051116	4.446	8.32 ± 0.18	291^{+80}_{-52}	11.3 ± 0.3	< 440	0.027 ± 0.020	0.077 ± 0.017
J020605.8−050441	4.497	5.98 ± 0.29	58^{+8}_{-6}	11.8 ± 0.3	< 460	0.166 ± 0.039	0.279 ± 0.035
J020611.7−050457	4.489	3.01 ± 0.22	111^{+70}_{-31}	8.6 ± 0.3	< 280	0.015 ± 0.021	0.073 ± 0.029
J020611.7−050627	4.460	1.39 ± 0.20	96^{+96}_{-33}	8.3 ± 0.3	< 260	0.044 ± 0.019	0.039 ± 0.022
J020614.1−050032	4.466	1.52 ± 0.20	22^{+4}_{-3}	7.7 ± 0.3	< 220	0.124 ± 0.028	0.186 ± 0.027
J142432.6+353825	4.428	4.33 ± 0.26	277^{+1648}_{-133}	7.4 ± 0.3	< 190	-0.011 ± 0.051	0.042 ± 0.057
J142439.8+353801	4.484	0.68 ± 0.42	> 16[h]	7.8 ± 0.3	< 230	-0.035 ± 0.059	-0.004 ± 0.060
J142545.5+352259	4.452	1.42 ± 0.09	110^{+37}_{-23}	10.3 ± 0.3	< 380	0.010 ± 0.012	0.035 ± 0.009
J142624.4+353832	4.457	2.71 ± 0.11	60^{+6}_{-6}	3.3 ± 0.3	< 270[g]	0.078 ± 0.013	0.122 ± 0.013

Continued on next page...

TABLE 5.1—*Continued*

Target	z^a	Lyα Flux[b]	$W_\lambda^{\text{rest c}}$ (Å)	FWHM[d] (Å)	Δv^e (km s^{-1})	Cont. (μJy)[f] Blue Side	Cont. (μJy)[f] Red Side
J142628.5+353808	4.407	6.54 ± 0.16	517^{+584}_{-179}	7.7 ± 0.3	< 220	-0.094 ± 0.011	0.034 ± 0.021

[a]The redshift was derived from the wavelength of the peak pixel in the observed line profile. We estimate the error in this measurement to be $0.002 < \delta_z < 0.004$, depending on the spectroscopic configuration. However, we note that this measurement may overestimate the true redshift of the system since the blue wing of the Lyα emission is absorbed by foreground neutral hydrogen.

[b]Units are 10^{-17} erg cm^{-2} s^{-1}. The line flux was determined by totaling the flux of the pixels that fall within the line profile. No attempt was made to model the emission line or to account for the very minor contribution of the continuum to the line. Quoted uncertainties account for photon counting errors alone, excluding possible systematic errors. Despite these caveats, the Lyα line fluxes measured from the spectra agree to 2σ in all but one case with those measured in the narrowband imaging.

[c]The rest frame equivalent widths were determined with $W_\lambda^{\text{rest}} = (F_\ell/f_{\lambda,r})/(1 + z)$, where F_ℓ is the flux in the emission line and $f_{\lambda,r}$ is the measured red–side continuum flux density. The error bars δw_+ and δw_- are 1σ confidence intervals determined by integrating over the probability density functions $P_i(w)$ described in § 5.4.2. The error bars are symmetric in probability density–space in the sense that $\int_{w-\delta w_-}^{w} P_i(w')\,dw' = \int_{w}^{w+\delta w_+} P_i(w')\,dw' = 0.34$.

[d]The FWHM was measured directly from the emission line by counting the number of pixels in the unsmoothed spectrum which exceed a flux equal to half the flux in the peak pixel. No attempt was made to account for the minor contribution of the continuum to the height of the peak pixel.

[e]The velocity width Δv was determined by subtracting in quadrature the effective instrumental resolution for a point source, and is therefore an upper limit, as the target may have angular size comparable to the $\lesssim 1''$ seeing of these data. Where the emission line is unresolved, the velocity width is an upper limit set by the effective width of the resolution element itself.

[f]Red and blue side continuum measurements are variance–weighted averages made in 1200 Å wide windows beginning 30 Å from the wavelength of the peak pixel in the emission line. A small correction factor was subtracted from the variance–weighted averages based on the detection of residual signal remaining in extractions of source–free, sky–subtracted regions of the two–dimensional spectra (see text, § 5.2.2). Quoted uncertainties account for photon counting errors in the source extractions added in quadrature to the photon counting errors derived in the blank–sky extractions.

[g]Line is unresolved.

[h]2σ lower limit. The measurement of the red–side continuum for this source is formally consistent with no observable flux. The equivalent width limit was then set by using a 2σ upper limit to $f_{\lambda,r}$ in the expression given in footnote (c).

Chapter 6

A Luminosity Function of Lyα–Emitting Galaxies at z ≈ 4.5

Abstract

We present a catalog of 59 $z \approx 4.5$ Lyα–emitting galaxies spectroscopically–confirmed in a campaign of Keck/DEIMOS follow–up observations to candidates selected in the Large Area Lyman Alpha (LALA) narrowband imaging survey. We targeted 97 candidates for spectroscopic follow–up; by accounting for the variety of conditions under which we performed spectroscopy, we estimate a selection reliability of $\sim 76\%$. Together with our previous sample of Keck/LRIS confirmations, the 59 sources confirmed herein bring the total catalog to 73 known $z \approx 4.5$ Lyα–emitting galaxies in the ~ 0.7 degrees^2 covered by the LALA imaging. As with the Keck/LRIS sample, we find that a significant fraction of the confirmed Lyα lines have rest–frame equivalent widths ($W_\lambda^{\mathrm{rest}}$) which exceed the maximum predicted for normal stellar populations: $17\% - 31\%$ of the detected galaxies show $W_\lambda^{\mathrm{rest}} > 190$ Å (93% confidence), and $12\% - 27\%$ show $W_\lambda^{\mathrm{rest}} > 240$ Å (90% confidence). Furthermore, we find that the Lyα lines systematically disfavor combinations of low velocity width (< 150 km s^{-1}) and high luminosity ($> 5 \times 10^{42}$ erg s^{-1}), suggesting that more than dust content alone, it is the velocity structure of a galaxy that determines the detectability of Lyα in emission. Finally, we construct a luminosity function of $z \approx 4.5$ Lyα emission lines for comparison to Lyα luminosity functions spanning $3.1 < z < 6.6$. We conclude that if there is evolution in the Lyα luminosity function over these epochs, its significance is below the statistical uncertainty of these data. This result supports the conclusion from several smaller samples of high–redshift Lyα–emitters that the intergalactic medium remains largely reionized from the local universe out to $z \approx 6.5$. However, it is somewhat at odds with the pronounced drop in the cosmic star formation rate density recently measured between $z \sim 3$ and $z \sim 6$ in Lyman–break galaxies, and therefore potentially sheds light on the relationship between the two populations.

6.1 Introduction

Observational cosmology has recently witnessed a tremendous increase in proficiency in the identification of galaxies at the earliest cosmic epochs. Thanks in large part to the availability of large–format mosaic CCDs well–suited for wide–field imaging and spectroscopic multiplexing, we are now transitioning from exotic, single detections of high–redshift galaxies (e.g. Dey et al. 1998; Weymann et al. 1998; Ellis et al. 2001; Ajiki et al. 2002; Dawson et al. 2002; Hu et al. 2002; Cuby et al. 2003; Taniguchi et al. 2003; Nagao et al. 2004; Rhoads et al. 2004; Stern et al. 2005) to the assembly of statistically robust samples spanning the earliest accessible redshifts. Robust samples of this kind are necessary for understanding the systematics of selection criteria, and of the spatial distribution of the galaxies themselves. Deficiencies therein are the main source of uncertainty in inferred luminosity functions and universal star formation rates, which in turn are the keys to understanding the cosmic history of star formation, galaxy assembly and evolution, and even the early ionization history of the IGM (e.g. Malhotra & Rhoads 2004; Stern et al. 2005).

Searches for high–redshift galaxies typically follow the by–now familiar strategy of targeting redshifted Lyα emission at increasing wavelengths with narrowband imaging in windows of low night–sky emission (e.g. Cowie & Hu 1998; Hu et al. 1998; Rhoads et al. 2000; Kodaira et al. 2003; Maier et al. 2003; Hu et al. 2004; Taniguchi et al. 2005), or by photometric selection in broadband imaging of the redshifted Lyman break (e.g. Steidel et al. 1996b; Madau et al. 1996; Lowenthal et al. 1997; Spinrad et al. 1998; Lehnert & Bremer 2003; Ando et al. 2004; Bouwens et al. 2004; Dickinson et al. 2004; Ouchi et al. 2004; Stanway et al. 2004a,b). These two techniques are complementary; Lyα searches at typical sensitivities can identify galaxies with UV–continua too faint to be detected by the Lyman break method, but such surveys only select that fraction of galaxies with strong line emission.

The Large Area Lyman Alpha (LALA) survey (Rhoads et al. 2000) has recently identified in deep narrowband imaging a large sample of Lyα–emitting galaxies at redshifts
$z \approx 4.5$ (Malhotra & Rhoads 2002), $z \approx 5.7$ (Rhoads & Malhotra 2001; Rhoads et al.
2003), and $z \approx 6.5$ (Rhoads et al. 2004). In Dawson et al. 2004 (Paper I), we reported on
the spectroscopic confirmation with Keck/LRIS of 17 Lyα–emitting galaxies selected in the
LALA $z \approx 4.5$ survey. The resulting sample of confirmed Lyα emission lines shows large
equivalent widths (median $W_\lambda^{\mathrm{rest}} \approx 80$ Å) but narrow velocity widths ($\Delta v < 500$ km s^{-1}),
indicating that the Lyα emission in these sources derives from star formation, not from
AGN activity. Moreover, though models of star formation in the primordial Universe predict that a small fraction of Lyα–emitting galaxies at $z \approx 4.5$ may be nascent, metal–free
objects (e.g. Scannapieco et al. 2003), and indeed we found with 90% confidence that 3 to
5 of the confirmed sources exceed the maximum Lyα equivalent width predicted for normal
stellar populations, we did not detect the He II λ1640 emission expected to be characteristic of primordial star formation. Specifically, the He II λ1640 flux in a composite of the 11
highest resolution spectra in the Keck/LRIS sample is formally consistent with zero, with
a 2σ (3σ) upper limit of 13% (20%) of the flux in the Lyα line. In other words, though
these galaxies may be young, they show no evidence of being truly primitive, Population
III objects.

We have recently more than quadrupled our catalog of spectroscopically–confirmed
Lyα–emitting galaxies at $z \approx 4.5$ with a campaign of Keck/DEIMOS spectroscopic follow–
up to candidates selected in the LALA survey. Together with the detections presented in
Paper I (and accounting for minor overlap in the samples), the 59 Lyα–emitters confirmed
with Keck/DEIMOS bring the total catalog to 73 known $z \approx 4.5$ Lyα–emitting galaxies in
the ~ 0.7 degrees^2 imaged by LALA. In this paper, we utilize these additional confirmations
to update the results of Paper I, and to construct a luminosity function of $z \approx 4.5$ Lyα
emission lines for comparison to Lyα luminosity functions spanning $3.1 < z < 6.6$. We
describe our imaging and spectroscopic observations in § 6.2, and we summarize the results
of the spectroscopic campaign in § 6.3. In § 6.4, we investigate the distribution in equivalent
width and in velocity width of the Lyα lines, we construct Lyα luminosity functions for our
sample and for several extant samples, and we discuss the implications of the luminosity
functions for the relationship between Lyα–emitters and Lyman–break galaxies (LBGs),
and for the history of reionization. We conclude in section § 6.5. Throughout this paper
we adopt a Λ–cosmology with $\Omega_{\mathrm{M}} = 0.3$ and $\Omega_\Lambda = 0.7$, and $H_0 = 70$ km s^{-1} Mpc^{-1}. At

$z = 4.5$, such a universe is 1.3 Gyr old, the lookback time is 90.2% of the total age of the Universe, and an angular size of $1''\!.0$ corresponds to 6.61 comoving kpc.

6.2   Observations

6.2.1   Narrowband and Broadband Imaging

The LALA survey concentrates on two primary fields, "Boötes" (14:25:57 +35:32; J2000.0) and "Cetus" (02:05:20 −04:55; J2000.0). Each field is 36 × 36 arcminutes in size, corresponding to a single field of the 8192 × 8192 pixel Mosaic CCD cameras on the 4m Mayall Telescope at Kitt Peak National Observatory and on the 4m Blanco Telescope at Cerro Tololo Inter–American Observatory. The $z \approx 4.5$ search uses five overlapping narrowband filters each with full width at half maximum (FWHM) ≈ 80 Å (Figure 6.1). The central wavelengths are 6559, 6611, 6650, 6692, and 6730 Å, giving a total redshift coverage of $4.37 < z < 4.57$ and a survey volume of 7.4×10^5 comoving Mpc^3 per field. In roughly 6 hours per filter per field, we achieve 5σ line detections in $2''\!.3$ apertures of $\approx 2 \times 10^{-17}$ erg cm^{-2} s^{-1}.

The primary LALA survey fields were chosen to lie within the NOAO Deep Wide–Field Survey (NDWFS; Jannuzi & Dey 1999). Thus, deep NDWFS broadband images are available in a custom B_W filter ($\lambda_0 = 4135$ Å, FWHM = 1278 Å; Jannuzi & Dey 1999, Jannuzi et al., in preparation) and in the Harris set Kron–Cousins R and I, as well as J, H, K, and K_s. The LALA Boötes field benefits from additional deep V and SDSS z' filter imaging. The imaging data reduction is described in Rhoads et al. (2000), and the $z \approx 4.5$ candidate selection is described in Rhoads & Malhotra (2001). Briefly, candidates are selected based on a 5σ detection in a narrowband filter, the flux density of which must be twice the R–band flux density, and must exceed the R–band flux density at the 4σ confidence level. To guard against foreground interlopers, we set a minimum observed equivalent width of $W_\lambda^{\mathrm{obs}} > 80$ Å, and the candidate must not be detected in the B_W–band.

6.2.2   Spectroscopic Observations

Between 2003 March and 2004 May we obtained spectroscopy of 97 $z \approx 4.5$ candidate Lyα–emitters with the Deep Imaging Multi–Object Spectrograph (DEIMOS; Faber et al. 2003), a next generation camera on the Keck II telescope with high multiplexing capabilities and improved red sensitivity. Each slitmask included approximately 15 candidate Lyα–

emitters (mixed in with roughly 50 other spectroscopic targets) and was observed for 1.5 to 2.0 hrs in 0.5 hr increments. Six slitmasks targeting a total of 80 candidates were observed in the Boötes field; the airmass in these observations never exceeded 1.5. One slitmask targeting 17 candidates was observed in the Cetus field; the airmass for this slitmask was constrained to less than 1.8. The seeing in all observations ranged from $0\rlap{.}''5$ to $1\rlap{.}''0$.

All observations employed $1\rlap{.}''0$ wide slitlets and the 600ZD grating ($\lambda_{\mathrm{blaze}} = 7500$ Å; 0.65 Å pixel^{-1} dispersion; $\Delta\lambda_{\mathrm{FWHM}} \approx 4.5$ Å ≈ 200 km s^{-1})[1]. No order–blocking filter was used; since the targets were primarily selected to have red colors, second order light should not be of concern. Most nights suffered from some cirrus; relative flux calibration was achieved from observations of standard stars from Massey & Gronwall (1990) observed during the same observing run. It should also be noted that the position angle of an observation was set by the desire to maximize the number of targets on a given slitmask, so observations were generally not made at the parallactic angle.

We processed the two–dimensional data using the DEEP2 DEIMOS pipeline[2]. We performed small ($0\rlap{.}''5$) dithers between exposures on our initial observing run; to reduce these data, we supplemented the DEEP2 DEIMOS pipeline with additional home–grown routines. We extracted spectra with the IRAF[3] package (Tody 1993) using the optimal extraction algorithm (Horne 1986), following standard slit spectroscopy procedures.

Prior experience with faint–object spectroscopy dictates that a small but significant error in the measured flux of faint continua may be introduced by sky subtraction during the processing of the two–dimensional spectra. We investigated this possibility with $\sim$ 10 additional non–overlapping extractions in source–free regions in each two–dimensional spectrum, parallel to and along the same trace as the extraction of the neighboring Lyα– emitting galaxy. We then fitted these blank–sky spectra over the same region that we fitted for the continuum redward and blueward of the emission line in the object extraction. For 15 sources, these fits yielded a tiny residual signal, which we interpreted as a systematic error in the two–dimensional sky subtraction and applied as a correction to quantities

[1] We measured the instrumental resolution by autocorrelating one–dimensional extracted spectra of night–sky emission lines. The autocorrelation results in an effective average line profile with a high signal–to–noise ratio, which we fit with a Gaussian to obtain the FWHM. We performed this test on $\sim$ 50 night–sky spectra with the result $\Delta\lambda_{\mathrm{FWHM}} = 4.47 \pm 0.03$ Å, where the error given is the error in the mean. Note that this error does not account for the possibility that some night–sky lines are blends, and so may slightly underestimate the uncertainty.

[2] See http://astron.berkeley.edu/~cooper/deep/spec2d/

[3] IRAF is distributed by the National Optical Astronomy Observatory, which is operated by the Association of Universities for Research in Astronomy, Inc., under cooperative agreement with the National Science Foundation.

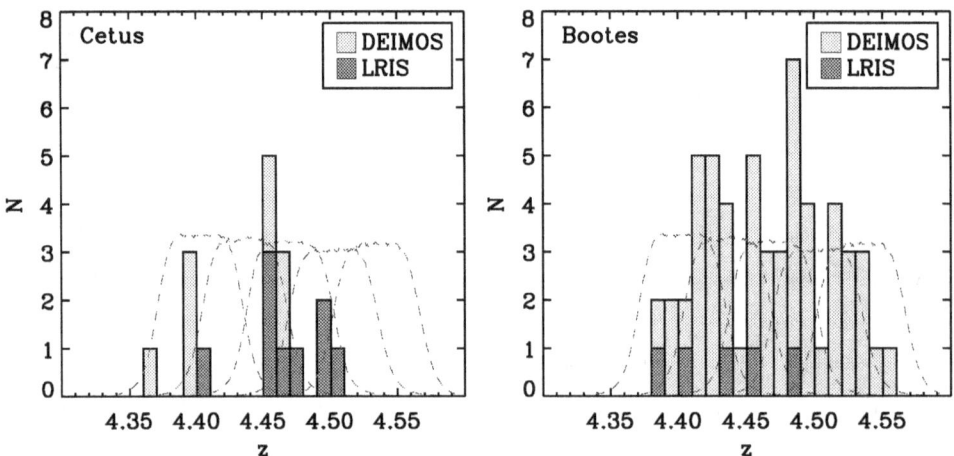

FIG. 6.1.— Distribution of redshifts for spectroscopically confirmed Lyα emission lines in the Cetus field (left; 02:05:20 −04:55, J2000.0) and in the Boötes field (right; 14:25:57 +35:32, J2000.0). The redshifts labeled "DEIMOS" denote galaxies confirmed with our campaign of Keck/DEIMOS spectroscopy, described in this paper. The redshifts labeled "LRIS" denote galaxies confirmed with our campaign of Keck/LRIS spectroscopy, described in Paper I. The overlays are arbitrarily scaled transmission curves for the five narrowband filters employed in the imaging component of this survey.

derived from the object spectra. The typical correction was $\sim 0.04 \pm 0.02$ μJy, but the correction reached as high as $\sim 0.1 \pm 0.07$ μJy in three cases. Sky–subtraction residuals of this kind generally resulted when a small spectroscopic slit contained a bright serendipitous detection in addition to the target, the combination of which made it difficult to fit the sky background.

6.3   Spectroscopic Results

Out of 97 spectroscopic candidates, we achieved 73 detections, 59 of which constitute Lyα confirmations according to the criteria outlined below. A histogram of these confirmations appears in Figure 6.1, and a set of sample spectra are shown in Figure 6.2. The spectroscopic properties of the Lyα confirmations are tabulated in Table 6.1. One detected galaxy lacks an emission line but shows a large spectral discontinuity identified as the onset of foreground Lyα–forest absorption at $z = 4.462$. Three of the detections are

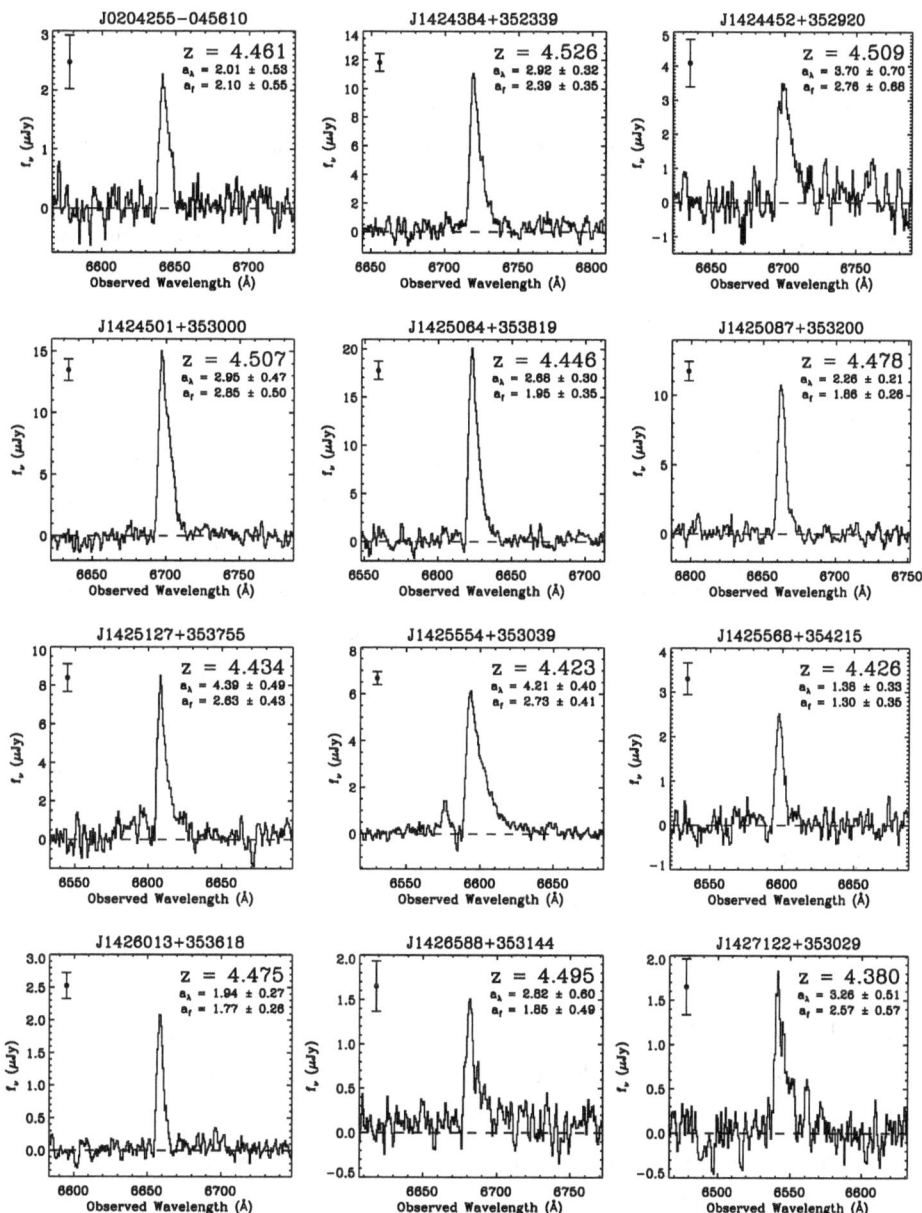

FIG.   6.2.— Sample spectra from the set of 59 $z \approx 4.5$ Lyα–emitting galaxies confirmed with Keck/DEIMOS, with a wavelength range selected to highlight the emission–line profile. The measured redshifts and asymmetry statistics (§ 6.3.2) are indicated in the upper right of each panel. The representative error bar (*upper left*) is the median of the flux error in each pixel over the wavelength range displayed. The spectra have been smoothed with a 3–pixel boxcar average.

identifiably low–redshift interlopers (two resolved [O II] λ3727 doublets at $z \sim 0.8$; one complex of [O III] λλ4959,5007 and Hβ at $z \sim 0.3$) which survived the candidate selection thanks to their unusually high equivalent widths (e.g. $W_\lambda^{obs} > 2000$ Å). Ten detections are of such low signal–to–noise ($\lesssim 1$) that they cannot be reliably identified as either Lyα or as low–redshift interlopers. If these 10 low signal–to–noise ratio detections could each be attributed to Lyα, then the "success rate" of the Keck/DEIMOS campaign would be 72%, identical to that of the Keck/LRIS sample described in Paper I (but also subject to all the caveats listed therein). We do not include unconfirmed sources in any of the ensuing discussion.

The remaining 24 targets were classified as nondetections. Five of these slitlets suffered from some kind of instrument or reduction issue, e.g. the target was dithered off the slitlet and so did not reproduce across the individual integrations, or irregularities in the machining of the slitmask resulted in defects in the data processing. Of the final 19 nondetections, 13 targets were observed under adverse conditions (e.g. variable cloud coverage, and/or poor seeing) for which the general spectroscopic yield was low. Our failure to confirm these targets as $z \approx 4.5$ Lyα–emitters should not be taken to bear on the efficacy of candidate selection.

Six nondetections were observed under photometric conditions with subarcsecond seeing for which the spectroscopic yield was otherwise high. However, subsequent inspection of the imaging revealed that five of these targets were suboptimal candidates for one of a variety of reasons: two candidates sit on a weak satellite residuals; one candidate appears in an initial epoch of imaging but not in subsequent epochs, suggesting that it is a variable source or a spurious detection; two candidates are marginal or irregular detections in the imaging. This leaves just one otherwise viable candidate Lyα–emitter that was not confirmed in spectroscopy, even though the conditions for spectroscopy were favorable. Since this source (J1424398+353801) was a single–band detection in the narrowband imaging, it is possible that it represents a spurious false–positive and not a genuine candidate. Given the large number ($10^{6.5}$) of independent resolution elements in the images, we expect about one false–positive at the 5σ–level per LALA field per narrowband filter, and this number could be larger if the noise properties of the image are not precisely Gaussian (see Rhoads et al. 2003).

To estimate the reliability of our candidate selection, we consider only the foregoing six non–detections, the three low–redshift interlopers, and the 10 low signal–to–noise ratio detections as legitimate non–confirmations. This admittedly rough scheme suggests a rate of 59 detections out of 78 viable candidates observed spectroscopically under workable conditions, for a final selection reliability of $\sim 76\%$. The rate of spectroscopic confirmation is plotted as a function of narrowband flux in Figure 6.3.

6.3.1   Spectroscopic Sensitivity to Lyα Emission

In principle, our spectroscopic sensitivity to Lyα emission of a given flux and at a given redshift could be estimated by determining the total sensitivity of the instrument, plus telescope, plus atmosphere at $(1216 \text{ Å})(1 + z)$ with observations of standard stars. However, as noted by Hogg et al. (1998), this process is complicated in practice by several effects. For instance, in a multislit observation, both the total wavelength coverage and the function mapping wavelength–to–pixels depend on the physical position of the source on the slitmask. Moreover, positional errors or slitmask alignment errors may result in random scatter in the precision to which sources are centered on their slitlets, leading to random scatter in throughput from source to source. Outside of instrumental effects, the sky brightness, color, and spectrum certainly vary from night to night, and often over the course of one night. Owing to these complications, we adapt the prescription given in Hogg et al. (1998) for a purely empirical approach to assessing sensitivity to line emission which employs the reduced spectra themselves.

Each one–dimensional spectrum was created with a variance–weighted optimal extraction (Horne 1986) from the two–dimensional data. For each object, we therefore have both the flux and the flux variance as a function of wavelength. We used the variance spectrum to estimate the uncertainties in quantities derived from the object spectra, e.g. we sum the variance spectrum in quadrature over the wavelength range covered by the observed line profile to estimate the uncertainty in the measured line flux. However, the variance may also be used to measure the noise over wavelength ranges corresponding to any Lyα line we *might* have detected given the redshift range permitted by our narrowband imaging, roughly $4.37 < z < 4.57$. Accordingly, for each object we ranged over redshift and calculated the smallest emission line flux detectable:

$$F_{\lim}(z) = n_{\text{sig}} \, \delta_{\text{disp}} \left\{ \sum_{\lambda = \lambda_1(z)}^{\lambda_2(z)} \sigma_\lambda^2 \right\}^{\frac{1}{2}} , \tag{6.1}$$

where n_{sig} is the minimum signal–to–noise ratio necessary for a detection (here taken to be 3), δ_{disp} is the grating dispersion (0.63 Å pix^{-1} for the Keck/DEIMOS 600ZD grating), and

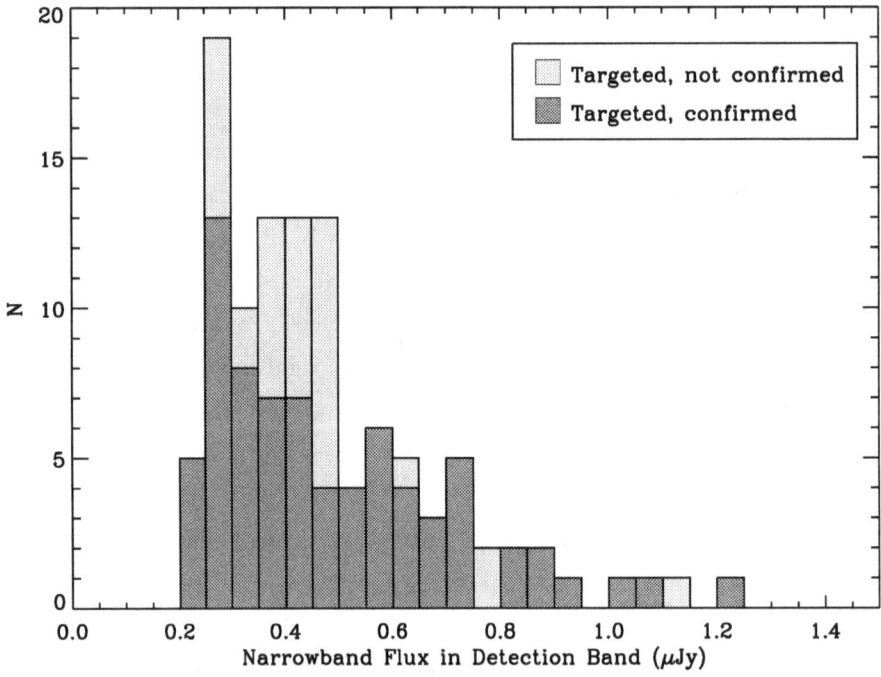

FIG. 6.3.— Spectroscopic success rate as a function of the flux in the narrowband in which the candidate was selected. This plot combines the results of the Keck/DEIMOS observations made for this paper and the Keck/LRIS observations described in Paper I.

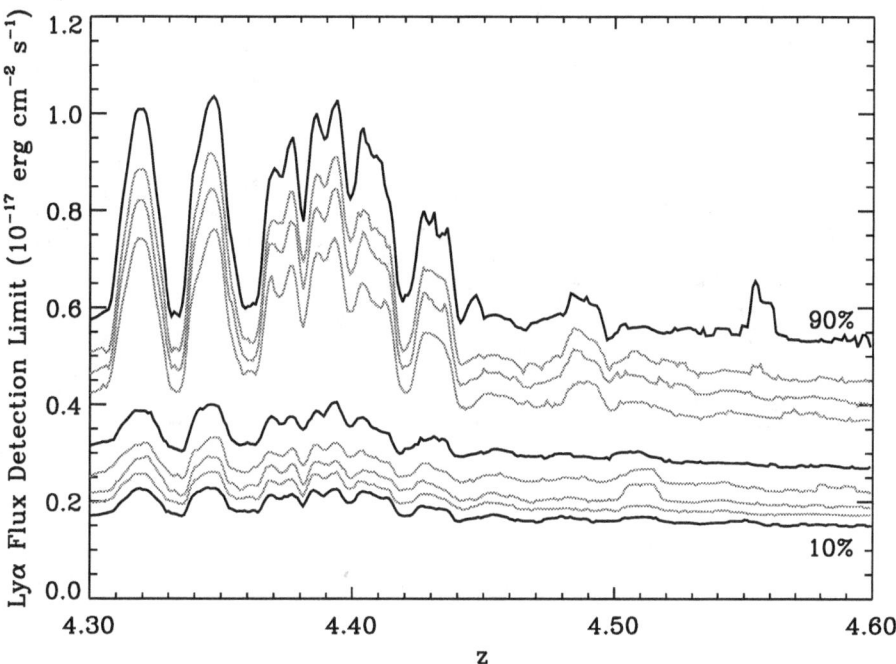

Fig. 6.4.— Empirical, cumulative distribution of spectroscopic sensitivity to Lyα emission, as a function of source redshift and Lyα flux. The contours span 10% to 90% in 10% steps. The dark lines denote the 10%, 50%, and 90% contours. The distribution is plotted cumulatively so that it can be interpreted as the probability that a putative Lyα emission line of a given flux and redshift would have been detected in our Keck/DEIMOS spectroscopic campaign.

σ_λ is the flux error in each pixel determined during the variance–weighted one–dimensional extraction, in units of f_λ. The limits λ_1 and λ_2 are defined by

$$\lambda_1(z) = (1+z)(1216 - \Delta\lambda/2)\,,$$
$$\lambda_2(z) = (1+z)(1216 + \Delta\lambda/2)\,, \tag{6.2}$$

where $\Delta\lambda$ is the fiducial rest–frame full width of the emission line (here taken to be 3 Å).

   We assembled the $F_{\mathrm{lim}}(z)$ for each object into a grid and then ranked the F_{lim} at each redshift, resulting in the cumulative distribution of sensitivity to Lyα emission line flux shown in Figure 6.4. The distribution may be interpreted as giving the probability that a putative Lyα emission line of a given flux and a given redshift would have been detected in

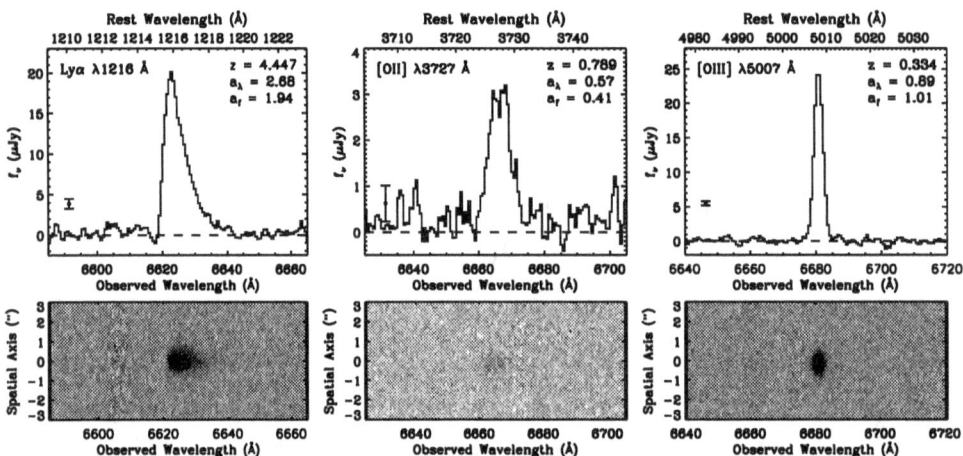

FIG. 6.5.— Sample Lyα emission line profile (*left*) compared to two common low–redshift interlopers: [O II] λ3727 (*center*) and [O III] λ5007 (*right*). The top figure in each case is the one–dimensional extracted spectrum; the bottom figure is a section of the two–dimensional data from which it was extracted. Note that we resolve the [O II] λ3727 doublet with our Keck/DEIMOS spectroscopic setup, thereby eliminating [O II] λ3727 as the main low–redshift interloper in our survey. The [O III] λ5007 line can typically be identified by neighboring [O III] λ4959 at one–third its strength, or by neighboring Hβ.

our spectroscopic campaign. Since we cover a comparatively small redshift range centered essentially at the peak of the detector throughput, the sensitivity distribution is dominated entirely by night–sky emission lines rather than by instrumental effects. And since the original narrowband survey was designed to probe relatively noise–free windows in night–sky emission, the spectroscopic sensitivity is fairly flat over the redshift range of interest. In sum, the implied depth of our spectroscopic survey is 50% complete to $f(Lyα) \sim 3^{-18}$ erg cm^{-2} s^{-1}, approximately 7 times deeper than the narrowband imaging. Note that because we derived this sensitivity function from the sample of spectra themselves, it depends entirely on the details governing the manner in which these spectra were obtained and processed, and is therefore valid for this survey only.

6.3.2   Redshift Identification

Of course, given the detection of an emission line, the identification of that line as high–redshift Lyα can remain problematic. Thorough treatments of the pitfalls of one–line redshift identifications are given elsewhere (e.g. Stern & Spinrad 1999; Stern et al. 2000a; Dawson et al. 2001). In surveys of the present kind, the primary threat to the proper interpretation of a solo mission line is the potential for low–redshift, high–equivalent width [O II] $\lambda3727$ to survive candidate selection, and then to be misidentified as high–redshift Lyα in later spectroscopy. However, at $z = 0.8$ (the redshift of an [O II] $\lambda3727$ line mistaken for Lyα at $z = 4.5$), the redshifted separation between the individual lines of the [O II] $\lambda3727$ doublet (rest wavelengths 3726 Å and 3729 Å, respectively) is 5.4 Å. The doublet is therefore just resolved in our spectroscopy and serves to uniquely flag [O II] $\lambda3727$ interlopers (Figure 6.5); this is an improvement afforded by Keck/DEIMOS over the spectroscopy presented in Paper I. Less frequently, high–equivalent width [O III] $\lambda5007$ survives as an interloper in our candidate selection. However, [O III] $\lambda5007$ can typically be identified by neighboring [O III] $\lambda4959$ at one–third its strength, or by neighboring Hβ.

Beyond merely eliminating plausible low–redshift interlopers, we may identify Lyα emission by its characteristically asymmetric morphology, and when the continuum is sufficiently well–detected, by the presence of a continuum break. Each of our confirmed Lyα detections demonstrates the asymmetric emission line profile characteristic of the line, where neutral hydrogen outflowing from an actively star–forming galaxy imposes a sharp blue cutoff and broad red wing (e.g. Dey et al. 1998; Stern & Spinrad 1999; Manning et al. 2000; Dawson et al. 2002; Rhoads et al. 2003; Hu et al. 2004; Stern et al. 2005; Taniguchi et al. 2005). In Figure 6.6, we present a scatter plot of the flux–based asymmetry statistic:

$$a_f = \frac{\int_{\lambda_p}^{\lambda_{10,r}} f_\lambda \, d\lambda}{\int_{\lambda_{10,b}}^{\lambda_p} f_\lambda \, d\lambda} \, , \tag{6.3}$$

versus the wavelength–based asymmetry statistic:

$$a_\lambda = \frac{(\lambda_{10,r} - \lambda_p)}{(\lambda_p - \lambda_{10,b})} \, , \tag{6.4}$$

for our sample, where λ_p is the wavelength of the peak of the emission line, and $\lambda_{10,b}$ and $\lambda_{10,r}$ are the wavelengths at which the line flux first exceeds 10% of the peak on the blue side and on the red side of the emission line, respectively (see Rhoads et al. 2003, 2004,

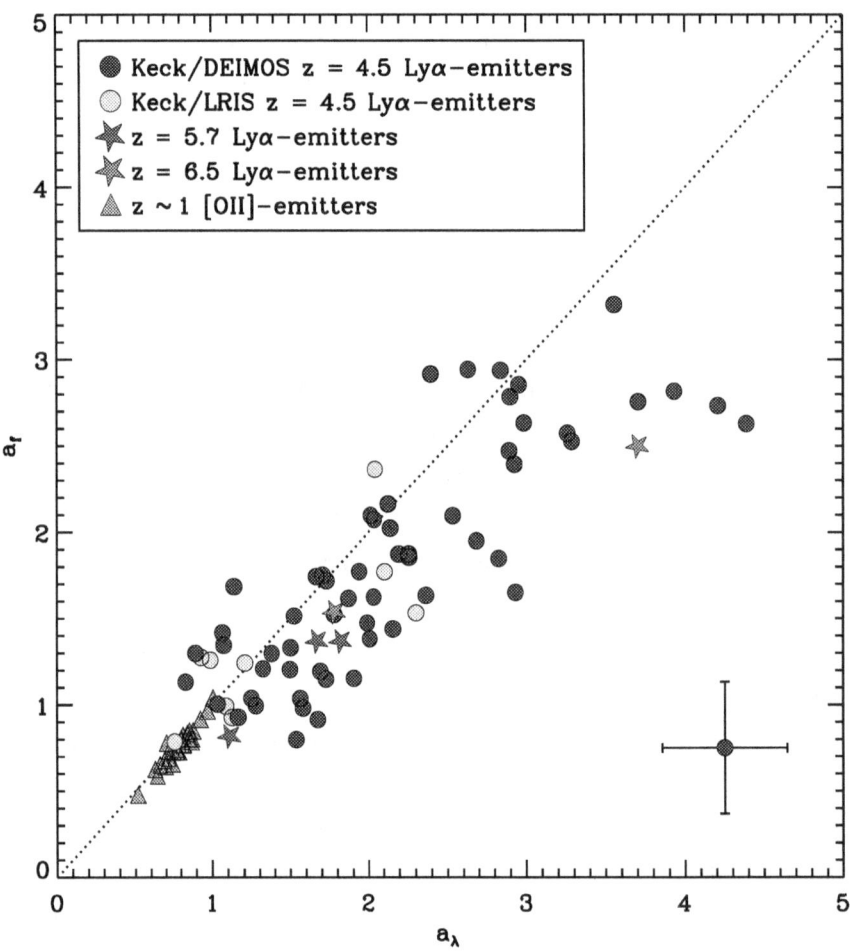

FIG. 6.6.— Scatter plot comparing the flux–based asymmetry statistic a_f and the wavelength–based asymmetry statistic a_λ of known high–redshift Lyα–emitters to a sample of [O II] $\lambda3727$–emitters at $z \sim 1$, updated from Paper I. The points labeled "DEIMOS" denote galaxies confirmed with our campaign of Keck/DEIMOS spectroscopy, described in this paper. The points labeled "LRIS" denote galaxies confirmed with our campaign of 400ℓ/mm–grating Keck/LRIS spectroscopy, described in Paper I. The three Lyα–emitters at $z = 5.7$ are from Rhoads et al. (2003), and the two Lyα–emitters at $z = 6.5$ are from Rhoads et al. (2004) and Stern et al. (2005), respectively. The 28 [O II] $\lambda3727$–emitters at $z \sim 1$ were provided by the DEEP2 team (Davis et al. 2003, A. Coil 2004, private communication); their Keck/DEIMOS 1200ℓ/mm–grating spectra were smoothed to the Keck/LRIS 400ℓ/mm–grating resolution by convolution with a Gaussian kernel. The representative error bar (*lower right*) is the median of the errors on the individual a_f and a_λ for the combined Keck/LRIS and Keck/DEIMOS sample.

and Paper I)[4]. Each of the confirmed Lyα emitters in this sample satisfies $a_f > 1.0$ or $a_\lambda > 1.0$, and 52 out of 59 sources satisfy both. Not surprisingly, as we found for the lower resolution Keck/LRIS sample in Paper I, the present Keck/DEIMOS sample of Lyα emitters is systematically segregated from low–redshift [O II] $\lambda3727$ in a_f–a_λ space when it is observed at resolutions typical of our Keck/LRIS and Keck/DEIMOS spectroscopic campaigns.

As a final diagnostic, we note that in each of our confirmed Lyα–emitters for which the continuum is sufficiently well–detected, the spectrum shows a continuum decrement consistent with the onset of absorption by the Lyα forest at $\lambda_{\rm rest} = 1216$ Å. The break amplitude is typically characterized by $1 - f_\nu^{\rm short}/f_\nu^{\rm long}$, where we define $f_\nu^{\rm short}$ as the variance–weighted flux density in a 1200 Å window beginning 30 Å below the emission line; $f_\nu^{\rm long}$ is the same, but above the emission line. In the 24 sources for which $f_\nu^{\rm long}$ is detected to better than 2σ, all but two sources have $1 - f_\nu^{\rm short}/f_\nu^{\rm long} > 0.5$, consistent with continuum break amplitudes at $z = 4.5$ in theoretical models (e.g. Madau 1995; Zhang et al. 1997), in the lower resolution Keck/LRIS sample presented in Paper I, and in other similar datasets (see Stern & Spinrad 1999, and references therein).

6.4   Discussion

Together with the observations presented in Paper I (and accounting for minor overlap in the samples), the 59 Lyα–emitters confirmed herein bring the total catalog of known $z \approx 4.5$ Lyα–emitting galaxies to 73 detections in the ~ 0.7 degrees^2 surveyed by the LALA imaging. We now update the characteristics of this population as they were estimated in Paper I by investigating the distribution of the total sample in both equivalent width and in velocity width. We then construct a $z \approx 4.5$ Lyα luminosity function, carefully accounting for survey incompleteness and for spectroscopic sensitivity, and we compare the result to Lyα luminosity functions spanning $3.1 < z < 6.6$.

---

[4] As in Paper I, the error bars on a_λ and a_f were determined with Monte Carlo simulations in which we modeled each emission line with the truncated Gaussian profile described in Hu et al. (2004) and Rhoads et al. (2004), added random noise in each pixel according to the photon counting errors, and measured the widths $\sigma(a_\lambda)$ and $\sigma(a_f)$ of the resulting distributions of a_λ and a_f for the given line. That is, for each $a_{\lambda,i}$, the error $\delta a_{\lambda,i} = \sigma(a_{\lambda,i})$, and similarly for each $a_{f,i}$.

6.4.1   The Equivalent Width and Velocity Width Distributions

As in Paper I, we determine the rest–frame equivalent widths directly from the spectra according to $W_\lambda^{\mathrm{rest}} = (F_\ell / f_{\lambda, r}) / (1 + z)$, where F_ℓ is the flux in the emission line and $f_{\lambda, r}$ is the measured red–side continuum flux density. The resulting equivalent width distribution is plotted in Figure 6.7, together with the equivalent widths measured in Paper I.

Before interpreting this distribution, one should be cautioned that the $W_\lambda^{\mathrm{rest}}$ determination is very sensitive to uncertainty in the measured continuum. Since the continuum estimate enters into the denominator of the expression for $W_\lambda^{\mathrm{rest}}$, the characteristically small continuum values and their large fractional uncertainties cause significant scatter in the measurement, and the resulting error is neither Gaussian nor symmetric about the measured value. Especially problematic is the fact that the largest values of $W_\lambda^{\mathrm{rest}}$ are also the least certain. Detailed discussions of the uncertainties in measuring $W_\lambda^{\mathrm{rest}}$ in high–redshift Lyα–emitters, along with the complicating effects of dust content, gas kinematics, and intergalactic absorption, are given in Hu et al. (2004) and in Paper I.

With these caveats in mind, we rigorously treated the error bars on the equivalent width estimates, and we restricted the analysis to sources with red–side continuum signal–to–noise ratios $\gtrsim 1$. To determine the equivalent width error bars, we first associated each measured line flux $F_{\ell,i} \pm \delta F_{\ell,i}$ with a Gaussian probability density function (PDF) centered on $F_{\ell,i}$ with width $\sigma = \delta F_{\ell,i}$; we proceeded similarly for the measured continuum fluxes. We then generated a grid of line flux versus continuum flux on which each node has an associated equivalent width and is assigned a weight given by the probability distribution on each of its flux axes. Next we collapsed the grid into a histogram of equivalent widths, adding the weight from each grid point to the appropriate equivalent width bin. The result is a non–Gaussian PDF $P_i(w)$ for which $P_i(w)\,dw$ is the probability of observing $W_{\lambda,i}^{\mathrm{rest}}$ in the interval $w < W_{\lambda,i}^{\mathrm{rest}} < w + dw$. The error bars δw_+ and δw_- are then 1σ confidence intervals determined by integrating over the probability density functions $P_i(w)$. They are symmetric in probability density–space in the sense that $\int_{w - \delta w_-}^{w} P_i(w')\,dw' = \int_{w}^{w + \delta w_+} P_i(w')\,dw' = 0.34$.

We find the resulting distribution to be broadly consistent with the equivalent widths presented in Fujita et al. (2003) for $z \sim 3.7$ and in Hu et al. (2004) for $z \sim 5.7$. While the majority of sources can be understood as comparatively young (1 to 10 Myr) galaxies with Salpeter initial mass functions (IMFs), a non–negligible fraction exceeds the largest rest–frame equivalent widths produceable by such stellar populations. Malhotra & Rhoads

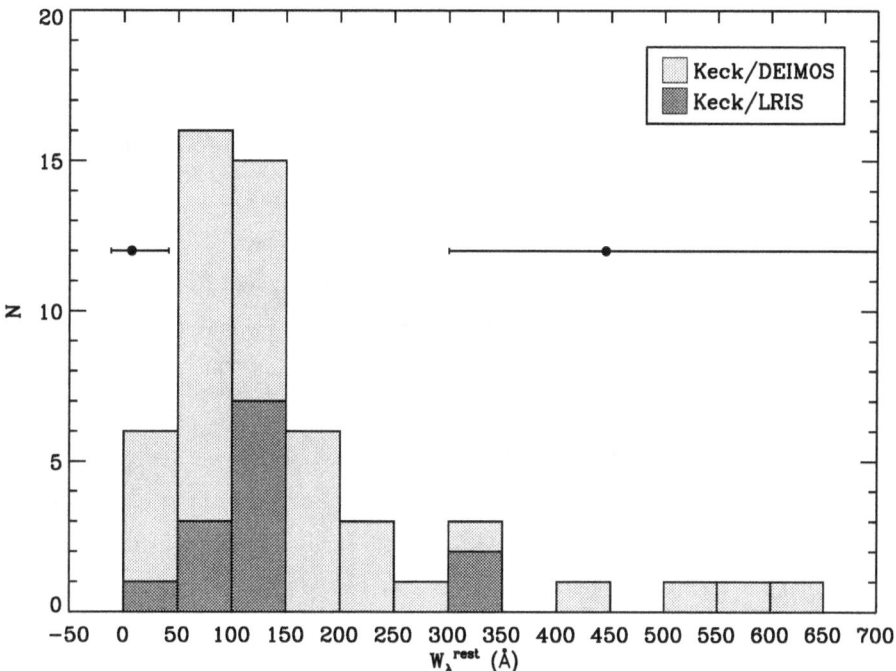

FIG. 6.7.— Histogram of the spectroscopic rest–frame equivalent widths for the $z \approx$ 4.5 population, determined with $W_\lambda^{\mathrm{rest}} = (F_\ell / f_{\lambda,r})/(1 + z)$, where F_ℓ is the flux in the emission line and $f_{\lambda,r}$ is the measured red–side continuum flux density. The sources labeled "DEIMOS" denote galaxies confirmed with our campaign of Keck/DEIMOS spectroscopy, described in this paper. The sources labeled "LRIS" denote galaxies confirmed with our campaign of Keck/LRIS spectroscopy, described in Paper I. Representative error bars on the equivalent widths are plotted at left and at right. Notably, the highest equivalent widths are generally the least certain, as they correspond to the faintest (and hence least certain) continuum estimates.

(2002) use a Salpeter initial mass function, an upper mass cutoff of 120 $M_\odot$, and a metallicity of 1/20th solar to find maximum Lyα equivalent widths of 300 Å, 150 Å, and 100 Å for stellar populations of ages 10^6, 10^7, and 10^8 years, respectively. Adopting a correction factor of 0.64 as an upper limit to the effect of IGM absorption on the measurement of $W_\lambda^{\mathrm{rest}}$ in spectroscopy effectively reduces these upper limits to 190 Å, 100 Å, and 60 Å(see discussion in Paper I). Owing to the lower metallicity used in their models, the pre–IGM–corrected values of Malhotra & Rhoads (2002) are slightly higher than the canonical limiting Lyα rest–frame equivalent width of 240 Å given by Charlot & Fall (1993).

Using the ensemble of $P_i(w)$ described above, we find that 12% – 27% of the galaxies in this sample show $W_\lambda^{\mathrm{rest}} > 240$ Å (90% confidence), and 17% – 31% show $W_\lambda^{\mathrm{rest}} > 190$ Å (93% confidence). Both results are nearly identical to the values given in Paper I. On the simplest interpretation, these galaxies are required to be very young (age $< 10^6$ years), or to have IMFs skewed in favor of the production of massive stars. The possibility that AGNs in our sample are producing stronger–than–expected Lyα emission seems unlikely due to the comparatively narrow velocity widths of the Lyα lines (see below) and to the absence of the high–ionization state UV emission lines symptomatic of AGN activity. Moreover, deep ($\sim$ 170 ks) *Chandra*/ACIS imaging of LALA $z \approx 4.5$ candidates in both Boötes (Malhotra et al. 2003) and in Cetus (Wang et al. 2004) resulted in X–ray non–detections to an average 3σ–limiting luminosity of $L_{2-8\mathrm{keV}} < 2.8 \times 10^{42}$ erg s^{-1}. This limit is roughly an order of magnitude fainter than what is typically observed for even the heavily obscured, Type II AGNs (e.g. Stern et al. 2002b; Norman et al. 2002; Dawson et al. 2003).

We also measured the velocity widths of the Lyα emission lines directly from the spectra, by subtracting in quadrature our estimated instrumental resolution from the FWHM of the line profile. We present a scatter plot of velocity width versus Lyα luminosity in Figure 6.8, and we include a selection curve depicting the minimum Lyα luminosity detectable for a given velocity width, determined from the measured noise per resolution element in a typical spectrum. Based on this selection effect alone, there is no *a priori* reason why combinations of low width (< 150 km s^{-1}) and high luminosity ($> 5 \times 10^{42}$ erg s^{-1}) should be systematically disfavored. The absence of objects in that region of parameter space is therefore suggestive of a physical effect, that a significant velocity gradient in the Lyα–emitting gas is a precondition on the escape of Lyα photons from the galaxy. Whereas the initial, largely failed attempts at the direct detection of high–redshift Lyα postulated that any production of metals and dust would quench Lyα emission (e.g. Djorgovski 1992;

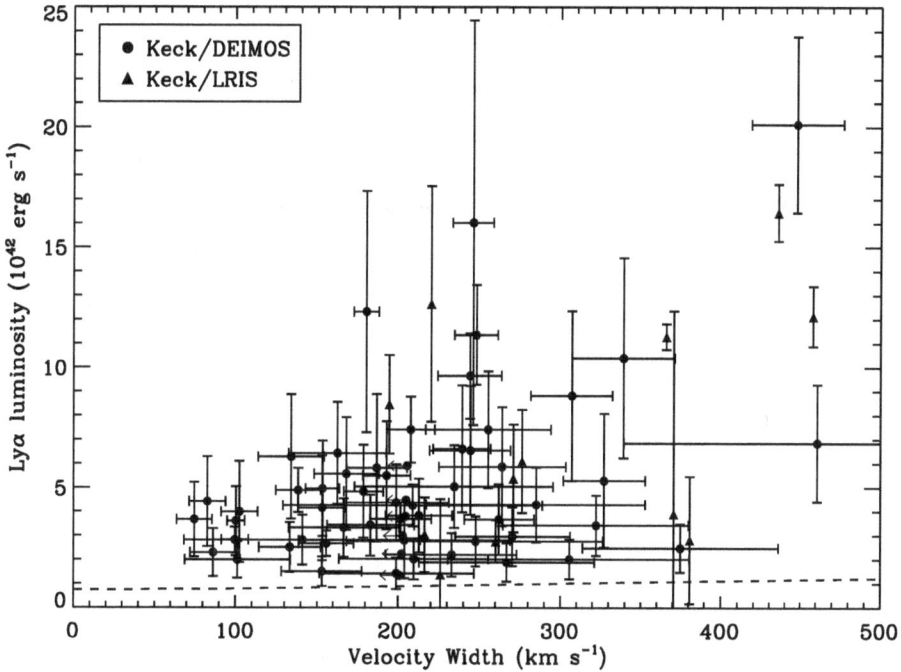

FIG. 6.8.— Scatter plot of velocity widths and luminosities of the Lyα emission lines for the galaxies reported in this paper (circles) and in Paper I (triangles). The dashed line depicts the minimum Lyα luminosity for a given velocity width, determined from the measured noise per pixel in a typical spectrum.

Pritchet 1994; Thompson & Djorgovski 1995), this result suggests that more than the dust content alone, it is the velocity structure of the galaxy which determines the detectability of Lyα in emission. By Doppler shifting the neutral hydrogen away from the Lyα line center, a velocity gradient in the gas has the effect of turning off resonant scattering, and thereby significantly diminishing a Lyα photon's pathlength to dust absorption.

Similar results have been reported both locally and at high redshift. Kunth et al. (1998) report that in the rest–frame UV spectra of 8 nearby star–forming galaxies only show Lyα in emission when the ISM metal lines are blue–shifted by ~ 200 km s^{-1} with respect to the ionized gas, though the sample spans a metal abundance of more than a factor of 10. Similarly, Lyman–break galaxies that show Lyα in emission typically show metallic ISM absorption lines which are blue–shifted by ~ 100 km s^{-1} (Pettini et al. 2001; Frye et al. 2002; Shapley et al. 2003). The emerging picture is that with or without the presence of dust, gas kinematics play a significant role in facilitating the escape of Lyα photons.

6.4.2 Empirical Cumulative Luminosity Function

In Figure 6.9, we present an empirical cumulative Lyα line luminosity function computed for our sample at $z \sim 4.5$ compared to luminosity functions computed for several other samples spanning $3.1 < z < 6.6$. The cumulative luminosity function gives for each Lyα line luminosity $L(\mathrm{Ly}\alpha)$ the total number density of Lyα lines brighter than $L(\mathrm{Ly}\alpha)$. The comparison samples are drawn from spectroscopic follow–up to narrowband surveys with roughly comparable flux limits and candidate–selection criteria (except where noted, below). In each case, we converted the reported Lyα line fluxes to line luminosities using a Λ–cosmology with $\Omega_M = 0.3$ and $\Omega_\Lambda = 0.7$, and $H_0 = 70$ km s^{-1} Mpc^{-1}, and we made a minimal attempt to account for incompleteness[5]. Specifically, the volume from which Lyα–emitting candidates were selected by their narrowband–excess is simply defined by the solid angle covered by the narrowband imaging, and by the redshift range allowed by the narrowband filter. However, the *effective* volume surveyed by the spectroscopic follow–up is smaller than the imaging survey volume by a factor of $N_{\mathrm{spec}}/N_{\mathrm{cand}}$, where N_{cand} is the total number of Lyα–emitting candidates discovered in the imaging, and N_{spec} is the number of candidates actually targeted for spectroscopy (e.g. Stern et al. 2005).

---

[5]Hu et al. (2004) provide fluxes for their $z \sim 5.7$ sources as measured in narrowband imaging, rather than Lyα line fluxes as measured in spectroscopy. As such, we adopt the conversion given in Stern et al. (2005) to estimate $f_{\mathrm{Ly}\alpha}$ from f_ν^{NB}.

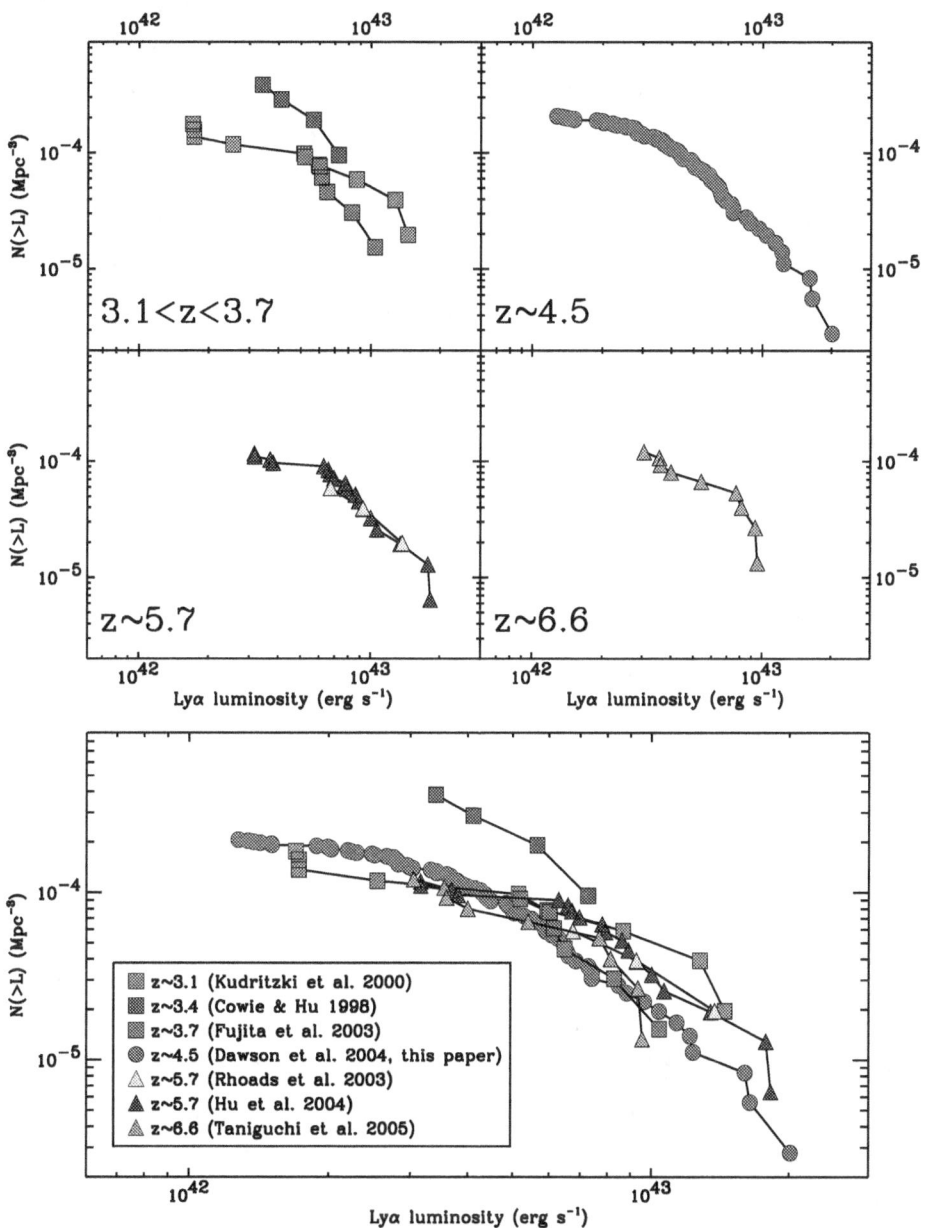

FIG. 6.9.— Comparison of empirical, cumulative Lyα luminosity functions computed with only minimal completeness correction for several spectroscopic surveys spanning $3.1 < z < 6.6$. The cumulative luminosity function gives for each Lyα line luminosity $L(\mathrm{Ly}\alpha)$ the total number density of Lyα lines brighter than $L(\mathrm{Ly}\alpha)$. No strong evolution is evident over the redshift range depicted.

No strong evolution is readily evident in the cumulative Lyα luminosity functions between $z \sim 3$ and $z \sim 6$. The only significant scatter between luminosity functions occurs between the $z \sim 3$ surveys, and that scatter likely finds its origin in differences in the manner in which the experiments were performed. Foremost, the area surveyed by the Cowie & Hu (1998) effort is comparatively small: just 25 arcmin^2 in each of two fields (HDF and SSA 22), as opposed to 300 arcmin^2 in Kudritzki et al. (2000) and 132 arcmin^2 in Fujita et al. (2003). Cowie & Hu (1998) note that the number counts in their HDF field appears to be 2.5 times richer in narrowband excess objects than their SSA 22 field, highlighting the susceptibility of small survey areas to cosmic variance. Separately, as noted by Hu et al. (2004), the Fujita et al. (2003) data may comparatively under–represent the density of Lyα–emitters due to their more stringent equivalent width criterion of $W_\lambda^{\rm obs} > 250$ Å, as opposed to $W_\lambda^{\rm obs} > 77$ Å in Cowie & Hu (1998) and effectively $W_\lambda^{\rm obs} \gtrsim 100$ Å in Kudritzki et al. (2000).

6.4.3   The $V/V_{\rm max}$ Estimate

To quantify these impressions, we now perform a more rigorous estimate of the Lyα luminosity functions with a modified version of the $V/V_{\rm max}$ method (e.g. Hogg et al. 1998; Fan et al. 2001). We investigate three redshift regimes by combining the three $z \sim 3$ surveys and the two $z \sim 5.7$ surveys with the $z \sim 6.6$ survey into single luminosity functions representing $z \sim 3$ and $z \sim 6$, respectively, for comparison with our data at $z \approx 4.5$. We then fit the luminosity function in each redshift regime with a Schechter function.

For each galaxy, $V_{\rm max}$ is the volume over which Lyα of a given luminosity could be located and still be detected by our survey; the luminosity function is then the sum of the inverse volumes of all galaxies in the given luminosity bins. Our modifications to the $V/V_{\rm max}$ method account for incompleteness in two senses. First, not every galaxy candidate identified in imaging was targeted in follow–up spectroscopy. Following Hogg et al. (1998), Figure 6.10 shows the fraction of narrowband–selected candidate Lyα–emitters which were targeted for spectroscopy as a function of flux in the band in which the candidate was detected. We label this *a priori* completeness function $\eta_{\rm try}$; the candidate Lyα flux $f_{\rm Ly\alpha}$ can be roughly estimated from the flux in the narrowband $f_\nu^{\rm NB}$ with $f_{\rm Ly\alpha} = w_n(f_\nu^{\rm NB} - f_\nu^{\rm R})$, where w_n is the width of the narrowband filter and $f_\nu^{\rm R}$ is the flux of the candidate in the R–band.

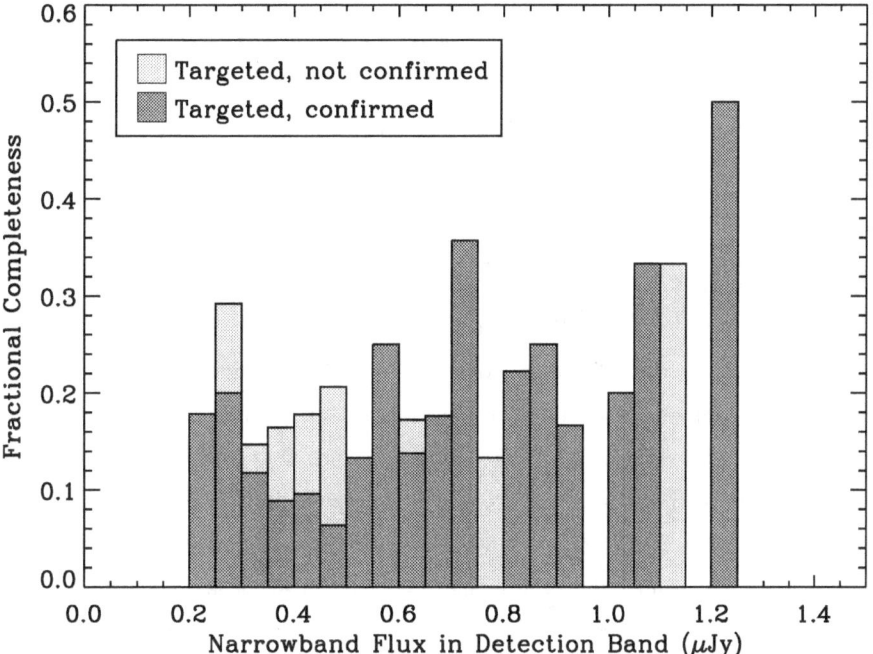

FIG. 6.10.— Probability as a function of narrowband flux that a candidate Lyα–emitter was targeted for spectroscopy, divided into the fraction of targets that were confirmed and the fraction of targets that were not.

Second, even if a candidate Lyα–emitter was selected for spectroscopy, its inclusion in the luminosity function depends on the detection and identification of the Lyα line. Our spectroscopic sensitivity to Lyα emission as a function of flux and redshift is shown in Figure 6.4; we label this function p_{detect}. As discussed in section § 6.3.1, p_{detect} can be interpreted as the probability that a putative Lyα emission line of a given flux and a given redshift would have been detected in our spectroscopic campaign.

In the presence of these selection effects, the available volume for a galaxy with Lyα emission of flux $f_{\text{Lyα}}$ is

$$V_{\max} = \int_{z_1}^{z_2} \eta_{\text{try}}(f'_{\text{Lyα}}) \, p_{\text{detect}}(f'_{\text{Lyα}}, z') \frac{d^2 V_c}{d\Omega \, dz'} \Delta\Omega \, dz' \,, \tag{6.5}$$

where the comoving volume element in a solid angle $d\Omega$ and redshift interval dz is the familiar

$$\frac{d^2 V_c}{d\Omega \, dz} = \left(\frac{c}{H_0}\right)^3 \left\{\int_0^z \frac{dz'}{E(z')}\right\}^2 \frac{1}{E(z)} \,, \tag{6.6}$$

with

$$E(z) = \left\{\Omega_{\text{M}}(1+z)^3 + \Omega_{\text{k}}(1+z)^2 + \Omega_\Lambda\right\}^{1/2} \,. \tag{6.7}$$

In equation 6.5, $\Delta\Omega$ is the solid angle covered by the LALA survey, and $f'_{\text{Lyα}}$ is the Lyα line flux for the source in question if it were located at redshift z'. The lower limit of integration z_1 is set by the lowest wavelength at which Lyα could be detected by our narrowband filters, corresponding to $z \approx 4.37$. The upper limit of integration z_2 is set in one of two ways. If the Lyα luminosity for a source is bright enough that the line remains above the survey flux limit out to the highest redshift accessible by our filter set, then z_2 is simply equal to the upper redshift limit for the survey, $z \approx 4.57$. For fainter sources, z_2 is taken to be the redshift at which the Lyα flux falls below the survey flux limit; in this case, $4.37 < z_2 < 4.57$.

We proceeded similarly when working with datasets provided by other authors, for which we generally did not have knowledge of the completeness or spectroscopic sensitivity to Lyα. In these cases, if the Lyα luminosity for a source was bright enough such that source remains above the stated survey flux limit out to the rear edge of the survey, then $V_{\max}$ is just the effective survey volume $V_{\text{eff}} = V_{\text{survey}}(N_{\text{spec}}/N_{\text{cand}})$ described in section § 6.4.2; that is, $V_{\max}$ is the stated total survey volume weighted by the fraction of candidates actually targeted for spectroscopy. For fainter sources, where the Lyα luminosity was such that the Lyα flux falls below the stated detection threshold at some redshift z_{detect} between z_1 and z_2, then $V_{\max}$ is just the weighted survey volume integrated from z_1 to z_{detect}.

Having computed V_{max} for each galaxy, we may compute the differential Lyα luminosity function $\Phi(L)$, the number density of galaxies per logarithmic interval in Lyα luminosity. In a given luminosity bin of width $\Delta \log L$ centered on L_i, this is given by

$$\Phi(L_i) = \frac{1}{\Delta \log L} \sum_j \frac{1}{V_{\mathrm{max},j}} \; . \tag{6.8}$$

Here, the index i denotes the luminosity bin and j denotes the galaxies within the bin, where the galaxies summed in a given bin are selected by their Lyα luminosities according to

$$|\log L_j - \log L_i| < \frac{\Delta \log L}{2} \; . \tag{6.9}$$

Finally, the uncertainty in the luminosity function may be estimated with

$$\sigma[\Phi(L_i)] = \frac{1}{\Delta \log L} \left[\sum_j \left(\frac{1}{V_{\mathrm{max},j}} \right)^2 \right]^{1/2} \; . \tag{6.10}$$

In Figure 6.11, we present Lyα luminosity functions in the redshift regimes $z \sim 3$, $z \sim 4.5$, and $z \sim 6$, and we fit the data in each regime with a Schechter function. If $\Phi(L)\,dL$ is the comoving number density of galaxies with luminosities in the range $(L, L+dL)$, then the corresponding Schechter function is

$$\Phi(L)\,dL = \frac{\Phi^*}{L^*} \left(\frac{L}{L^*} \right)^\alpha \exp\left(-\frac{L}{L^*} \right) dL \; , \tag{6.11}$$

where Φ^* is the normalization, L^* is the characteristic break luminosity, and α sets the slope at the faint end. This is related to the number density of galaxies in *logarithmic* intervals by

$$\Phi(L)\,d(\log L) = \left(\frac{L}{\log_{10} e} \right) \left(\frac{\Phi^*}{L^*} \right) \left(\frac{L}{L^*} \right)^\alpha$$
$$\exp\left(-\frac{L}{L^*} \right) d(\log L) \; , \tag{6.12}$$

and it is this function which we fit to our data. As in van Breukelen et al. (2005), because the binned data points are few, we choose to fix $\alpha = -1.6$ so as to fit with only two free parameters, Φ^* and L^*. This choice fits well with the luminosity distribution of both LBGs and Lyα–emitters at $z \approx 3$ (Steidel et al. 1999, 2000). We determine the best fit with the Levenberg–Marquardt technique for solving the nonlinear least–squares problem (Press et al. 1992); the derived parameters are tabulated in Table 6.2.

With the exception of Φ^* for the $z \sim 3$ fit, the derived parameters agree to 1σ across redshift regimes. Moreover, our luminosity functions are in good agreement with

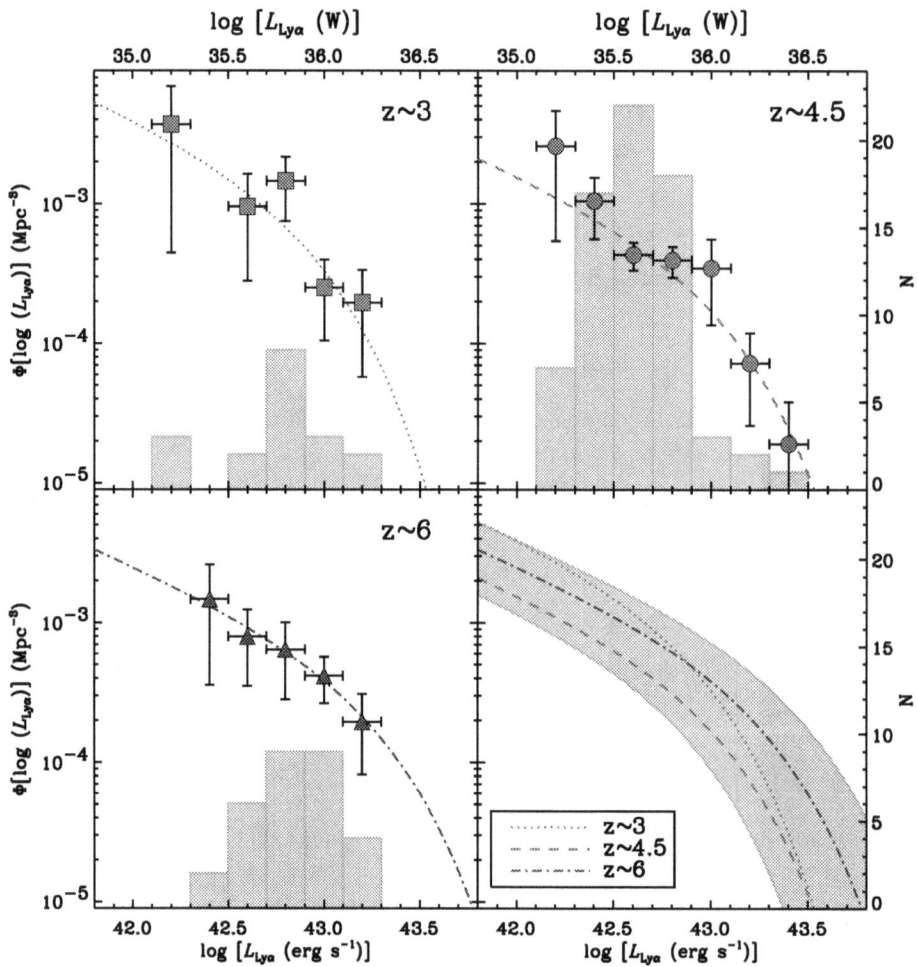

FIG. 6.11.— Lyα luminosity functions computed with the $V/V_{\max}$ method for several spectroscopic surveys spanning $3.1 < z < 6.6$, and fit with Schechter functions. The $z \sim 3$ bin combines the surveys presented by Kudritzki et al. (2000), Cowie & Hu (1998), and Fujita et al. (2003). The $z \sim 4.5$ bin combines the Keck/DEIMOS data presented in this paper and the Keck/LRIS data presented in Paper I. The $z \sim 6$ bin combines the surveys presented by Rhoads et al. (2003), Hu et al. (2004), and Taniguchi et al. (2005). For comparison, just the Schechter functions are overlaid in lower right–hand panel. The grey shaded area depicts the area covered by the 2σ error bars on the $z \sim 6$ Schechter function fit parameters.

those generated for samples compiled with alternative methods, e.g. for a sample of Lyα–emitters spanning $2.3 < z < 4.6$ discovered in integral–field spectroscopy (van Breukelen et al. 2005), and for a sample of $z \approx 5.7$ Lyα–emitters selected in the vicinity of the quasar SDSSp J104433.04−012522.2 (Ajiki et al. 2003). For these reasons, we interpret the slightly larger scatter in the $z \sim 3$ determination of Φ^* as symptomatic of the systematics already discussed (§ 6.4.2), and we conclude that if there is evolution in the Lyα luminosity function between $z \approx 3$ and $z \approx 6$, its significance is below the statistical uncertainty of these data.

6.4.4   Comparison to LBGs

At first glance, this result might be interpreted to corroborate the observed non–evolution of the rest–UV luminosity density and cosmic star formation rate (SFR) derived from LBGs over the same redshift range. Estimates of the $z \approx 6$ SFR based on the Great Observatories Origins Deep Survey/Advanced Camera for Surveys (GOODS/ACS; Giavalisco et al. 2004a) imaging of the Hubble Deep Field (HDF) suggest that star formation activity among the LBGs decreases only mildly with redshift, being $\sim 25\%$ depressed from a peak occurring between redshifts $1 < z < 3$ (Giavalisco et al. 2004b). This conclusion is in good agreement with similar measurements from both *HST* and ground–based datasets (e.g. Bouwens et al. 2003; Lehnert & Bremer 2003; Iwata et al. 2003, but see Stanway et al. 2003), and suggests that the onset of substantial cosmic star formation must have taken place earlier than $z > 6$. As the Lyα luminosity of a galaxy is directly related to its SFR (e.g. Dey et al. 1998; van Breukelen et al. 2005), the simple interpretation of the non–evolution of Lyα luminosity functions is the same as that for the LBGs: that we have not yet sampled sufficiently early epochs to witness a turn–over in cosmic star formation.

Subsequent measurements of the $z \approx 6$ SFR from LBGs selected in the recently-released *HST*/ACS Hubble Ultra–Deep Field (UDF) have complicated this simple interpretation. Bunker et al. (2004) point out that prior efforts based on shallower datasets show substantial disagreement in the derived surface densities of LBGs at $z \approx 6$, ranging from 0.1 arcmin^{-2} (Stanway et al. 2003) to 2.3 arcmin^{-2} (Yan et al. 2003). This disagreement could arguably be attributed to cosmic variance, but is more likely owed at least in part to the fact that the characteristic luminosity L^* of LBGs corresponds to a challenging z–band magnitude of $z'_{AB} \sim 26$ by redshifts $z \gtrsim 6$. Probing sufficiently faint along the LBG luminosity function to reliably include the contribution of galaxies less luminous than L^* therefore requires observers to work at low signal–to–noise ratios, where the tendency is for more foreground objects with intrinsically bluer colors to scatter up into the color–selection than for real LBGs to scatter out through photometric errors. The exceptionally deep imaging comprising the UDF allows the same faint galaxies to be probed at higher signal–to–noise. Two independent analyses of galaxy candidates in the UDF (Bunker et al. 2004; Yan & Windhorst 2004) have resulted in nearly identical catalogs of $z \approx 6$ LBGs, with the result that the cosmic SFR measured at $z \approx 6$ is approximately 6 times less than that at $z \approx 3$.

What does this development mean for the Lyα–selected galaxies? Steidel et al. (2000) report that Lyα–selection with an equivalent width criterion typical of narrowband surveys would return 20% – 25% of their $z \approx 3$ LBGs. If Lyα–emitters are merely a subset of the LBG population which happen to have been detected during a stage of strong Lyα production, than we would expect the ensemble of Lyα–emitters to show similarly depressed star formation activity by $z \approx 6$. Integrating the luminosity functions assembled from our data and assuming a one–to–one relationship between Lyα luminosity density and cosmic star formation density suggests $\rho_{\mathrm{SFR}}(z \sim 6)/\rho_{\mathrm{SFR}}(z \sim 3) = 0.8 \pm 0.7$. Clearly, this is not a strong constraint. However, we may speculate that the Lyα–emitters as a population are evolving differently from the LBGs. No measurements of stellar mass based on observations yet exist for Lyα–emitters, but Lyα–selected galaxies have lower bolometric luminosities than LBGs at the same redshift, and should therefore be less massive. Assuming a monotonic relationship between galaxy and halo mass (e.g. Mo et al. 1998), Lyα–emitting galaxies should be found in smaller, more numerous dark matter halos, and as a population may plausibly have a different evolutionary history from that of the LBGs.

Alternatively, it remains possible that assembling increasingly robust datasets of Lyα–emitters at fainter fluxes and higher redshifts will reveal the same depressed Lyα SFR as was measured in the UDF for LBGs. The resolution of this issue, as well as the disentangling of the LBG and Lyα–emitting populations, will clearly rely on further observations. In addition to more and deeper spectroscopy, a coordinated campaign of space–based near– and mid–infrared would provide the rest–frame optical measurement necessary to constrain the ages, masses, and metallicities of the galaxies' stellar populations (e.g. Yan et al. 2005, Mobasher et al., in progress).

6.4.5   Implications for Reionization

The spectroscopic observations of the $z > 6$ quasars yielded the first detections of the long–awaited Gunn–Peterson trough, seeming to imply that the intergalactic medium (IGM) is neutral from $z \approx 1000$ to $z \approx 6$ (Becker et al. 2001; Djorgovski et al. 2001; Fan et al. 2002). Subsequently, the *Wilkinson Microwave Anisotropy Probe* (WMAP) identified a large amplitude signal in the temperature–polarization maps of the cosmic microwave background (Spergel et al. 2003) indicating a large optical depth to Thomson scattering and favoring reionization instead at $z \approx 15$. The WMAP results are not necessarily inconsistent with those of the quasar Gunn–Peterson troughs, however. As summarized by Stern et al. (2005), only a small neutral fraction ($x_{\mathrm{HI}}^{\mathrm{IGM}} \sim 0.001$) is required to produce the Gunn–Peterson effect, so one plausible scenario is that reionization was an extended event, beginning early but not completing until $z \approx 6$. Alternatively, a variety of theoretical models now suggest that reionization occurred twice, first at $z \approx 20$ with the onset of zero–metallicity Population III stars, and then again by massive Population II stars formed after a partial recombination (e.g. Cen 2003; Haiman & Holder 2003; Somerville et al. 2003).

High–redshift Lyα–emitting galaxies offer another angle of attack on this issue, as the visibility of Lyα emission should be a sensitive function of the IGM neutral fraction (e.g. Santos 2004). Stern et al. (2005) and Malhotra & Rhoads (2004) present first attempts to exploit this fact by comparing luminosity functions of Lyα–emitters at $z \sim 5.7$ and $z \sim 6.6$. Both studies find no measurable evolution between these epochs, from which they infer that the IGM remains largely reionized from the local universe out to $z \approx 6.5$. Thanks in large part to the nine additional $z \approx 6.6$ sources recently provided by Taniguchi et al. (2005), the sample presented herein represents a three–fold increase in size over the prior efforts, and attains the same null result.

Significantly, the related question of *what* is responsible for reionization remains at large. It has long been recognized that AGN at early epochs are insufficient, owing to their rapid decline in space density at high redshift (e.g. Madau et al. 1999; Barger et al. 2003). Based on their analysis of the UDF, Bunker et al. (2004) conclude that the cosmic SFR produced by $z \approx 6$ LBGs is roughly five times too low to reionize the Universe, even in the unlikely event that the escape fraction of ionizing photons is unity. We estimate that the contribution to the cosmic SFR from Lyα–emitters at this epoch is lower than that of the LBGs ($\rho_{\mathrm{SFR}}(\mathrm{Ly}\alpha) \approx 0.003\ M_\odot\ \mathrm{yr}^{-1}\ \mathrm{Mpc}^{-3}$, as compared to $\rho_{\mathrm{SFR}}(\mathrm{LBG}) \approx 0.005\ M_\odot\ \mathrm{yr}^{-1}\ \mathrm{Mpc}^{-3}$) when integrated over the same limits. Consequently, though high–redshift

Lyα–emitters are proving to be a useful probe of the history of reionization, they are evidently not its cause. The implication is again that the bulk of reionization occurred much earlier than $z \approx 6$, or, less likely, that a significant source of ionizing photons remains unaccounted for.

6.5   Conclusion

Together with the observations reported in Paper I, the spectroscopic confirmations presented herein constitute a total sample of 73 Lyα–emitting galaxies at $z \approx 4.5$, representing the most comprehensive set of spectroscopic follow–up to narrowband–selected candidates at any redshift. As in Paper I, a significant fraction of the 59 Lyα emission lines confirmed with Keck/DEIMOS show rest–frame equivalent widths which exceed the maximum value predicted for normal stellar populations; the simplest interpretation is that these galaxies are required to be very young (age $< 10^6$ years), or to have top–heavy IMFs. The lack of high–ionization state UV emission lines confirms that the high–equivalent width Lyα emission is not powered by AGN activity. Rather, the tendency of the lines to avoid combinations of low velocity width and high luminosity suggests that a velocity gradient in the emitting gas is a precondition on the escape of Lyα photons.

Additionally, when we compare the luminosity function of Lyα–emitters at $z \approx 4.5$ to luminosity functions for similarly assembled samples spanning $3.1 < z < 6.6$, we find no evidence for evolution over these epochs. This result bolsters the conclusion from smaller catalogs of high–redshift Lyα–emitting galaxies assembled by Malhotra & Rhoads (2004) and Stern et al. (2005) that the IGM remains largely reionized from the local universe out to $z \approx 6.5$. However, it is somewhat at odds with the factor of six drop in the cosmic star formation rate density measured by Bunker et al. (2004) between $z \sim 3$ and $z \sim 6$ in Lyman–break galaxies selected in the exceptional imaging of the UDF. Admittedly, the constraint to the cosmic SFR we provide is much weaker than that derived from the UDF. Nonetheless, it is possible that this result indicates that these two populations follow different evolutionary histories. In either case, the disentanglement of this issue will likely rely on spaced–based near– and mid–infrared imaging of the high–redshift Lyα–emitters. Such observations, targeting the galaxies' rest–frame optical, would finally provide direct access to the heretofore unmeasured ages, masses, and metallicities of their stellar populations.

Acknowledgements

This work benefited greatly from conversations with M. Cooper, S. McCarthy, T. Robishaw, and J. Simon. In addition, we are humbly indebted to the expert staff of W. M. Keck Observatory for their assistance in obtaining the data herein. It is a pleasure to thank P. Amico, J. Lyke, and especially G. Wirth for their invaluable assistance during observing runs. Finally, we wish to acknowledge the significant cultural role that the summit of Mauna Kea plays within the indigenous Hawaiian community; we are fortunate to have the opportunity to conduct observations from this mountain. The work of D. S. was carried out at the Jet Propulsion Laboratory, California Institute of Technology, under contract with NASA. A. D. and B. J. acknowledge support from NOAO, which is operated by the Association of Universities for Research in Astronomy, Inc., under cooperative agreement with the National Science Foundation (NSF). H. S. gratefully acknowledges NSF grant AST 95–28536 for supporting much of the research presented herein. This work made use of NASA's Astrophysics Data System Abstract Service.

TABLE 6.1

SPECTROSCOPIC PROPERTIES OF THE KECK/DEIMOS $z \approx 4.5$ Lyα–EMITTERS

Target	z^{a}	Lyα Flux[b]	$W_{\lambda}^{rest\,c}$ (Å)	FWHM[d] (Å)	Δv^{e}	Cont. $(\mu Jy)^{f}$ Blue Side	Cont. $(\mu Jy)^{f}$ Red Side
J020418.2−050748	4.449	2.55 ± 0.87	$> 86^{g}$	6.8 ± 1.7	230	-0.040 ± 0.026	-0.018 ± 0.049
J020423.2−050647	4.449	3.25 ± 1.07	$> 108^{g}$	5.7 ± 1.0	160	-0.003 ± 0.026	0.014 ± 0.034
J020425.5−045610	4.461	3.72 ± 1.22	379^{+2092}_{-187}	7.2 ± 1.0	260	0.002 ± 0.025	0.026 ± 0.031
J020425.7−045810	4.387	1.98 ± 0.68	$> 39^{g}$	4.1 ± 0.6	$< 200^{h}$	-0.035 ± 0.052	0.021 ± 0.057
J020427.4−050045	4.390	1.47 ± 0.54	$> 142^{g}$	5.4 ± 2.8	140	-0.010 ± 0.018	-0.011 ± 0.019
J020428.5−045924	4.390	3.57 ± 1.27	508^{+4493}_{-278}	11.0 ± 2.8	460	-0.008 ± 0.032	0.019 ± 0.033
J020429.8−050251	4.460	1.39 ± 0.52	$> 22^{g}$	7.1 ± 2.2	250	-0.124 ± 0.070	0.012 ± 0.077
J020432.3−045519	4.360	3.13 ± 1.04	$> 241^{g}$	4.2 ± 1.3	$< 210^{h}$	-0.003 ± 0.023	-0.009 ± 0.022
J142434.9+352833	4.423	1.13 ± 0.46	26^{+30}_{-12}	6.8 ± 1.1	230	0.037 ± 0.077	0.117 ± 0.073
J142436.0+352600	4.464	1.81 ± 0.72	38^{+17}_{-16}	5.0 ± 0.2	100	0.016 ± 0.022	0.128 ± 0.025
J142438.4+352339	4.526	3.21 ± 1.28	26^{+11}_{-10}	7.0 ± 0.5	240	0.175 ± 0.036	0.336 ± 0.061
J142445.2+352920	4.509	1.21 ± 0.49	9^{+4}_{-3}	9.5 ± 1.5	370	0.106 ± 0.041	0.350 ± 0.049
J142445.3+352450	4.475	2.47 ± 0.99	$> 55^{g}$	5.6 ± 0.3	150	-0.056 ± 0.043	-0.009 ± 0.065
J142445.4+352859	4.514	0.98 ± 0.40	6^{+2}_{-2}	8.2 ± 2.0	310	0.174 ± 0.040	0.447 ± 0.052
J142450.1+353000	4.507	4.32 ± 1.73	83^{+70}_{-36}	8.2 ± 0.6	310	-0.010 ± 0.050	0.141 ± 0.063
J142452.4+352613	4.411	1.97 ± 0.79	97^{+212}_{-46}	6.5 ± 0.6	210	0.051 ± 0.038	0.054 ± 0.047
J142458.6+353558	4.522	2.02 ± 1.06	$> 24^{g}$	5.6 ± 0.5	150	-0.026 ± 0.078	-0.006 ± 0.116
J142459.8+353927	4.482	1.98 ± 1.05	$> 59^{g}$	5.0 ± 0.5	100	0.030 ± 0.041	-0.009 ± 0.049
J142501.7+353652	4.496	1.38 ± 0.78	43^{+721}_{-24}	6.4 ± 2.0	200	0.073 ± 0.100	0.088 ± 0.140
J142502.8+353017	4.476	0.75 ± 0.31	$> 20^{g}$	5.6 ± 0.9	150	0.052 ± 0.034	0.016 ± 0.042
J142503.4+353222	4.489	0.66 ± 0.28	21^{+18}_{-9}	4.1 ± 2.2	$< 200^{h}$	0.113 ± 0.037	0.086 ± 0.044
J142506.4+353819	4.446	8.11 ± 4.26	594^{+4407}_{-336}	7.0 ± 0.3	250	0.008 ± 0.054	0.037 ± 0.065
J142508.3+353952	4.511	2.59 ± 1.36	175^{+844}_{-93}	8.6 ± 0.6	330	-0.019 ± 0.038	0.040 ± 0.048
J142508.7+353200	4.478	2.41 ± 0.96	$> 75^{g}$	6.0 ± 0.4	180	0.032 ± 0.040	-0.010 ± 0.049
J142512.0+353913	4.451	1.13 ± 0.60	$> 30^{g}$	4.1 ± 1.4	$< 200^{h}$	0.063 ± 0.040	0.000 ± 0.050
J142512.7+353755	4.434	2.96 ± 1.56	34^{+19}_{-17}	6.1 ± 0.6	190	0.201 ± 0.040	0.235 ± 0.053
J142518.0+353415	4.408	5.37 ± 2.15	39^{+18}_{-15}	8.7 ± 0.8	340	0.150 ± 0.062	0.370 ± 0.072
J142522.4+353553	4.519	1.79 ± 0.72	39^{+30}_{-16}	7.4 ± 0.6	260	0.011 ± 0.046	0.126 ± 0.053
J142525.9+352349	4.471	3.27 ± 1.34	33^{+16}_{-13}	7.0 ± 0.7	240	0.072 ± 0.048	0.267 ± 0.057

Continued on next page...

TABLE 6.1—*Continued*

Target	z^a	Lyα Flux[b]	$W_\lambda^{\mathrm{rest}\,c}$ (Å)	FWHM[d] (Å)	Δv^e	Cont. (μJy)[f] Blue Side	Cont. (μJy)[f] Red Side
J142526.2+352531	4.464	2.76 ± 1.13	85^{+105}_{-39}	6.2 ± 0.4	190	0.067 ± 0.050	0.087 ± 0.054
J142531.8+352652	4.482	0.94 ± 0.40	18^{+9}_{-7}	7.4 ± 1.5	270	0.034 ± 0.035	0.140 ± 0.041
J142532.9+353013	4.534	5.49 ± 1.00	201^{+75}_{-51}	7.1 ± 0.3	250	0.005 ± 0.015	0.075 ± 0.018
J142535.2+352743	4.449	6.23 ± 2.54	159^{+173}_{-72}	6.0 ± 0.2	180	0.001 ± 0.048	0.106 ± 0.057
J142539.5+353902	4.432	1.52 ± 0.67	240^{+2182}_{-126}	4.0 ± 1.6	< 200[h]	0.049 ± 0.019	0.017 ± 0.022
J142541.7+353351	4.409	3.24 ± 1.34	108^{+139}_{-50}	5.4 ± 0.8	130	-0.042 ± 0.045	0.080 ± 0.050
J142542.0+352557	4.393	1.05 ± 0.44	30^{+28}_{-13}	6.4 ± 1.4	210	0.028 ± 0.033	0.092 ± 0.042
J142542.6+352626	4.450	1.49 ± 0.62	19^{+10}_{-7}	7.5 ± 0.9	270	0.101 ± 0.043	0.215 ± 0.059
J142544.5+354325	4.533	2.84 ± 1.20	131^{+129}_{-60}	7.4 ± 1.1	260	0.002 ± 0.018	0.059 ± 0.030
J142546.8+354315	4.443	0.72 ± 0.33	40^{+34}_{-19}	6.3 ± 1.5	200	0.012 ± 0.016	0.049 ± 0.022
J142547.8+354200	4.539	1.11 ± 0.48	$> 56^g$	4.9 ± 0.8	90	0.029 ± 0.017	0.007 ± 0.024
J142548.4+352740	4.546	1.21 ± 0.50	$> 24^g$	5.4 ± 0.7	130	-0.058 ± 0.038	0.004 ± 0.067
J142555.4+353039	4.423	10.31 ± 1.88	560^{+467}_{-190}	10.8 ± 0.6	450	0.025 ± 0.015	0.049 ± 0.023
J142556.7+354234	4.425	2.26 ± 0.96	189^{+284}_{-89}	4.8 ± 0.6	80	-0.019 ± 0.020	0.032 ± 0.022
J142556.8+354215	4.426	2.85 ± 1.20	162^{+152}_{-74}	5.8 ± 0.7	170	0.021 ± 0.020	0.047 ± 0.022
J142559.8+353513	4.394	1.39 ± 0.28	158^{+189}_{-60}	5.6 ± 0.6	160	0.005 ± 0.013	0.023 ± 0.014
J142559.8+353748	4.420	4.95 ± 0.91	55^{+10}_{-10}	7.0 ± 0.5	240	0.109 ± 0.018	0.240 ± 0.019
J142601.3+353618	4.475	1.41 ± 0.27	$> 171^g$	5.0 ± 0.4	100	-0.013 ± 0.011	-0.003 ± 0.013
J142602.0+354554	4.473	1.83 ± 0.78	85^{+70}_{-38}	4.8 ± 0.7	70	-0.003 ± 0.019	0.058 ± 0.025
J142612.2+353541	4.418	1.90 ± 0.36	$> 140^g$	6.3 ± 0.5	200	0.082 ± 0.013	0.002 ± 0.017
J142624.4+353832	4.460	2.46 ± 0.46	320^{+1182}_{-145}	5.4 ± 0.5	140	-0.012 ± 0.017	0.021 ± 0.022
J142627.5+353717	4.488	2.11 ± 0.43	42^{+24}_{-14}	6.5 ± 2.4	210	0.044 ± 0.025	0.135 ± 0.051
J142628.5+353809	4.409	3.82 ± 0.71	$> 143^g$	6.4 ± 0.4	210	-0.038 ± 0.029	-0.011 ± 0.041
J142653.5+353356	4.494	2.12 ± 0.77	142^{+115}_{-59}	7.8 ± 1.8	280	-0.016 ± 0.015	0.040 ± 0.017
J142658.8+353144	4.495	2.16 ± 0.79	31^{+11}_{-10}	6.3 ± 1.0	200	-0.019 ± 0.015	0.191 ± 0.018
J142706.3+353224	4.480	0.99 ± 0.38	100^{+139}_{-45}	5.0 ± 1.6	100	0.021 ± 0.014	0.027 ± 0.018
J142709.1+352738	4.407	1.77 ± 0.65	85^{+50}_{-34}	8.4 ± 1.4	320	0.039 ± 0.014	0.056 ± 0.017
J142709.2+352409	4.520	1.62 ± 0.59	$> 85^g$	5.8 ± 1.2	170	0.029 ± 0.015	0.008 ± 0.022
J142709.8+352641	4.405	1.78 ± 0.66	178^{+337}_{-82}	6.0 ± 0.8	180	0.023 ± 0.017	0.027 ± 0.021

Continued on next page...

Table 6.1—*Continued*

Target	z^{a}	Lyα Flux^{b}	$W_\lambda^{\mathrm{rest\,c}}$ (Å)	FWHM^{d} (Å)	Δv^{e}	Cont. $(\mu\mathrm{Jy})^{\mathrm{f}}$ Blue Side	Cont. $(\mu\mathrm{Jy})^{\mathrm{f}}$ Red Side
J142712.2+353029	4.380	2.35 ± 0.86	82^{+45}_{-32}	4.3 ± 2.4	$< 200^{\mathrm{h}}$	0.035 ± 0.017	0.076 ± 0.021

[a]The redshift was derived from the wavelength of the peak pixel in the observed line profile smoothed with a 3–pixel boxcar average. We estimate the error in this measurement to be $\delta_z \approx 0.0005$, based on Monte Carlo simulations in which we added random noise to each pixel of every spectrum according to the photon counting statistics, and then re–measured the redshift in each case. We note that this measurement may overestimate the true redshift of the system since the blue wing of the Lyα emission is absorbed by foreground neutral hydrogen.

[b]Units are 10^{-17} erg cm^{-2} s^{-1}. The line flux was determined by totaling the flux of the pixels that fall within the line profile. No attempt was made to model the emission line or to account for the very minor contribution of the continuum to the line. Quoted uncertainties account for photon counting errors alone, excluding possible systematic errors. Despite these caveats, the Lyα line fluxes measured from the spectra agree to 1σ in all but three cases with those measured in the narrow band imaging.

[c]The rest frame equivalent widths and their error bars were determined as described in § 6.4.1.

[d]The FWHM was measured directly from the emission line by counting the number of pixels in the unsmoothed spectrum which exceed a flux equal to half the flux in the peak pixel. No attempt was made to account for the minor contribution of the continuum to the height of the peak pixel.

[e]Units are km s^{-1}. The velocity width Δv was determined by subtracting in quadrature the effective instrumental resolution for a point source, and is therefore an upper limit, as the target may have angular size comparable to the $\lesssim 1''$ seeing of these data. Where the emission line is unresolved, the velocity width is an upper limit set by the effective width of the resolution element itself.

[f]Red and blue side continuum measurements are variance–weighted averages made in 1200 Å wide windows beginning 30 Å from the wavelength of the peak pixel in the emission line. We employed a 10–iteration, 2σ clipping algorithm to reduce the effect of spurious outliers occurring at long wavelength, where the sky noise is large. In some cases, a small correction factor was subtracted from the variance–weighted averages based on the detection of residual signal remaining in extractions of source–free, sky–subtracted regions of the two–dimensional spectra (see text, § 6.2.2). Quoted uncertainties account for photon counting errors in the source extractions added in quadrature to the photon counting errors derived in the blank–sky extractions.

[g]2σ lower limit. The measurement of the red–side continuum for this source is formally consistent with no observable flux. The equivalent width limit was then set by using a 2σ upper limit to $f_{\lambda,r}$ in the expression given in footnote (c).

[h]This line is unresolved.

TABLE 6.2

SCHECHTER FUNCTION FIT PARAMETERS

Redshift Bin	α	L^* $(10^{42}$ erg s$^{-1})$	Φ^* $(10^{-4}$ Mpc$^{-3})$
$z \sim 3$	-1.6	8.4 ± 5.7	5.2 ± 1.6
$z \sim 4.5$	-1.6	10.9 ± 3.3	1.7 ± 0.2
$z \sim 6$	-1.6	18.5 ± 5.4	2.0 ± 0.1

NOTE.—The error bars on L^* and Φ^* are the 1σ formal errors computed from the covariance matrix in the nonlinear least–squares fit, scaled by the measured value of χ^2. That is, $\delta L^* = \sigma_{L^*}\sqrt{\chi^2/n_{\mathrm{DOF}}}$, and similarly for $\delta\Phi^*$ (Press et al. 1992).

Chapter 7

Conclusion

As late as 1995, the anticipated widespread population of primeval galaxies remained at large, lurking undetected at unknown redshifts, with undiscovered properties. Only a handful of high–redshift Lyα–emitting galaxies had been detected (e.g. Djorgovski et al. 1985, 1987; Hu & McMahon 1996; Hu et al. 1996; Petitjean et al. 1996), and these sources had been found exclusively in the vicinity of objects already known to be at high redshift. As such, the detections were discounted as a possible consequence of the anomalous environments around the target objects, and were therefore not interpreted as the long–awaited signature of primeval galaxies in their primary epoch of formation.

Thanks largely to improvements in telescope aperture size and in detector area and sensitivity, the intervening decade has witnessed tremendous success in finally detecting this elusive primeval population. From the inaugural successes with narrowband imaging selection (e.g. Cowie & Hu 1998), to the dramatic first discovery of a galaxy at $z > 5$ in serendipitous spectroscopy (Dey et al. 1998), observational cosmology has transitioned away from exotic, single detections of high–redshift galaxies (e.g. Weymann et al. 1998; Ellis et al. 2001; Ajiki et al. 2002; Dawson et al. 2002; Hu et al. 2002; Cuby et al. 2003; Taniguchi et al. 2003; Nagao et al. 2004; Rhoads et al. 2004; Stern et al. 2005) to the assembly of statistically robust samples spanning the earliest accessible redshifts. This thesis presents a snapshot of precisely this transition. The campaign of serendipitous slit spectroscopy chronicled in Chapters 2, 3, and 4 yielded a significant fraction of the handful of high–redshift galaxies then known, and provided among the first suggestions of their ensemble properties. Chapters 5 and 6 describe the implementation of narrowband imaging selection, with which we traded redshift coverage for survey volume, focusing on the systematic study of galaxies at a particular epoch rather than chasing that rare, most–distant beast.

This thesis therefore provides one account of the manner by which observational cosmology has shifted from merely marveling at the incredible lookback times implied by the first few high–redshift detections, to the routine assembly of high–redshift datasets designed to address specific astrophysical issues. To conclude, we now elucidate some of these issues, and we outline observational programs which would lead to their resolution.

7.1   Connecting Galaxy Populations

Both high–redshift Lyα–emitting galaxies and high–redshift Lyman–break galaxies (LBGs) are selected based on their properties in the rest–frame UV. At best, our observed–frame optical detections provide access to the measurement of a single emission line (Lyα), and in some cases, to a measurement of the rest–frame UV continuum. Consequently, very little is known about the rest–frame optical properties of high–redshift galaxies. SEDs for the Lyman–break galaxy population are just now beginning to emerge, thanks primarily to new mid–infrared data obtained with the *Spitzer Space Telescope* as part of its in–orbit checkout activities. These early results are somewhat at odds. *Spitzer*/Infrared Array Camera (IRAC) observations of a sample of $z \sim 3$ LBGs in the vicinity of the bright QSO HS 1700+6416 suggest that the previous optical/near–infrared studies of LBGs did not miss large, hidden old stellar populations which, if present, would have been too faint in the rest–frame UV to be detected in the presence of younger stars[1] (Barmby et al. 2004). On the other hand, *Spitzer* and *HST* observations of the $z \sim 7$ galaxy strongly lensed by the massive galaxy cluster A2218 are consistent with a significant Balmer break, suggesting that a mature stellar population is already in place, even at such a high redshift (Egami et al. 2005). Clearly, more deep infrared observations of LBGs are merited.

In contrast to the burgeoning, though still small, suite of mid–infrared observations of LBGs, *no* serious attempt has been made to study the Lyα–emitting galaxies at these wavelengths. Most fundamentally, this lack of knowledge of their rest–frame optical properties means that we have little knowledge of their stellar populations, such that we cannot yet map galaxy samples to each other across observed redshifts. More specifically, there is growing evidence that the rest–frame UV data does not tell the whole story concerning

---

[1] Such a population has been hypothesized to exist in the Lyα–emitting galaxies in an effort to explain the exceptionally high rest–frame Lyα equivalent widths sometimes observed; see § 7.2.

these sources. First, as we discussed in Chapter 5, many submillimeter–selected galaxies at high redshift show Lyα in emission (Chapman et al. 2003), indicating that Lyα emission may emerge from systems with substantial dust masses which would not be obvious in studies of their rest–frame UV. Second, the spatial correlation length for the Lyα–emitting population is comparatively high, implying that these galaxies reside in massive dark matter halos which should in turn host more baryons than we have yet seen. We expand on both of these themes, below.

7.2   The Physical Nature of the Lyα–Emitting Galaxies

The observations in this thesis have borne out the now 38–year old prediction that copious ionizing radiation from hot, young stars interacting with primordial or near–primordial interstellar gas should produce strong Lyα signatures in young galaxies (Partridge & Peebles 1967). We have detected Lyα–emitting galaxies spanning $2.5 \lesssim z \lesssim 6$; the known galaxies at $z > 5$ are nearly all Lyα–emitters, and yet very few Lyα–emitters are known at $z = 0$. Based on this circumstantial evidence, plus the measurements of large rest–frame equivalent widths not easily accounted for by normal stellar populations, can we conclude that Lyα–emitters are primarily nascent galaxies found at high redshifts, and that we are seeing their evolution?

Indeed, as discussed in Chapters 5 and 6, the large rest–frame equivalent widths seen in $\sim 20\%$ of the detected Lyα emission lines could potentially be the product of young ($t < 10^7$ years), metal–poor ($Z = 0.05\,Z_\odot$) stellar populations in their first bursts of star formation. Stellar populations older than 10^7 years result in substantially lower equivalent widths, as they raise the production of UV continuum photons without increasing the production of line photons. Furthermore, the narrow velocity widths of the Lyα lines, along with the lack of high–ionization state emission lines and the non–detection of the sources in deep *Chandra* X–ray imaging (Malhotra et al. 2003; Wang et al. 2004) rule out AGN activity as an alternative culprit. For these reasons, the conclusion that Lyα–emitters are young galaxies in a young universe seems credible.

Several lines of evidence suggest that the foregoing argument may be overly simplistic. Foremost, we did not detect the signature He II $\lambda 1640$ emission in either individual or composite spectra (to a 2σ [3σ] upper limit of 13% [20%] of the flux in the Lyα line), effectively ruling out the youngest stellar population models (e.g. Schaerer 2003). Moreover, Shapley et al. (2001, 2003) argue that the strongest Lyα emission lines observed in the $z \sim 3$ LBG sample stem from the oldest stellar populations there observed. Additionally, Stiavelli et al. (2001) report on a sample of Lyα–selected galaxies at $z \sim 2.4$ whose continuum colors suggest a background old (> 1 Gyr) stellar population.

Finally, perhaps most damning to the "Lyα emission is tantamount to a young stellar population" conclusion is the fact that the stellar population models which produce the high equivalent width emission generally fail to consider the influence of dust and/or kinematics on the radiation transfer. The detailed geometry of the galaxy's interstellar medium (ISM) doubtless plays several roles in this process, any number of which could be acting to facilitate the creation and emission of high equivalent width Lyα. One such scenario is radiative transfer through a clumpy, dusty ISM. Dust in a *homogeneous* ISM effectively quenches Lyα emission (e.g. Meier & Terlevich 1981), thanks to the long pathlength to dust absorption caused by resonant scattering in the static, neutral hydrogen. By contrast, a clumpy ISM, where the dust and neutral gas have a high covering factor but a low volume filling factor, acts to suppresses continuum radiation through strong absorption in the high–density regions, while line photons resonantly scatter in the low–density regions, ultimately making their escape (Neufeld 1991). The result is a high equivalent width emission line.

In addition to the quantity and distribution of dust, there is mounting evidence that the kinematics of the emitting regions are also a strong determinant of the strength and nature of the Lyα emission. Lyα–emitting galaxies at low redshift tend to be galaxies with strong winds (e.g. Kunth et al. 1998), though the emission lines are two orders of magnitude less luminous than those observed at high redshift. Nonetheless, typical models of star–forming galaxies result in an overpressured cavity of hot gas inside the galaxy, which ultimately expands and bursts out into the halo in the form of a galactic wind (e.g. Tenorio–Tagle et al. 1999; Heckman et al. 2000; Mas–Hesse et al. 2003). The wind acts to Doppler shift the absorbers, minimizing resonant scattering of Lyα line of photons, thereby decreasing their pathlength to dust absorption and facilitating their escape. Meanwhile, dust continues to absorb continuum photons, and once again, the result is a high equivalent width emission line.

If such a kinematical scenario were indeed a precondition on Lyα emission, then we would expect to see a positive correlation between Lyα luminosity and Lyα velocity width. Indeed, we see some evidence of such a correlation in local galaxies (Kunth et al. 1998), as well as in the $z \approx 4.5$ population described in Chapters 5 and 6. On the other hand, no sign of extended line emission due to winds was found in a recent study of 46 unresolved source candidates in the Hubble Ultra Deep Field (Pirzkal et al. 2005), though such outflows could easily be less extensive than the 0.6 kiloparsec resolution, or could fall below the surface brightness threshold of the imaging.

The resolution of this issue, as well as those in the foregoing section, will clearly rely on further observations. Deeper spectroscopy at higher resolution could shed light on the morphologies of the Lyα–emission lines, offering the sort of insight into the relationship between velocity width and Lyα emission we explored in Chapter 3. Moreover, stacked spectra could potentially reveal velocity structure via comparison of ISM absorption lines to the stellar rest–frame of the galaxy. An even more optimal scenario would be a coordinated campaign of imaging with *HST* in the near–infrared and *Spitzer* in the mid–infrared. Both an old stellar population or that of a dusty starburst would leave a signature in *Spitzer*/IRAC 3.6 μm observations. To differentiate between them, *HST*/NICMOS H and Gemini/NIRI K observations could be made to target the Balmer spectral break at $\lambda_{\mathrm{rest}} = 4000$ Å. The SED of a dusty starburst would be smooth across that break; the SED of an older stellar population would show a break that is highly pronounced. Alternatively — and perhaps most tantalizingly — is the scenario in which the Lyα–emitting galaxies remain undetected at 3.6 μm. Such a nondetection, given an observation of sufficient depth, would imply no dust *and* no old stellar population, confirming that the Lyα–emitters are indeed young, free of dust, and undergoing their first bursts of star formation.

7.3 Host Dark Matter Halos

It is not yet clear how the high–redshift Lyα–emitting galaxies fit into the over–all story of the hierarchical formation of structure. The mass spectrum and merger histories of the dark matter halos which play host to luminous galaxies are determined by gravity alone, and thus can be predicted from first principles. However, the onset and evolution of star formation within a given dark matter halo is a complicated phenomenon. The simplest paradigm relating the two suggests a monotonic relationship between halo and galaxy mass (e.g. Mo, Mao, & White 1998). If so, Lyα–selected galaxies, with bolometric luminosities lower than those of LBGs at similar redshifts, should be found in smaller, more numerous dark matter halos showing weaker spatial correlation. However, the measured correlation strength of the Lyα–selected galaxies is comparable to that of the LBGs (with a correlation length of $r_0 \approx 6$ Mpc). To reconcile this discrepancy, we hypothesize that the correlation strength of the Lyα–selected galaxies at small radii is dominated by cases where multiple Lyα–emitting galaxies occupy the same dark matter halo, thereby increasing the counts of close pairs (e.g. Bullock et al. 2002).

Alternatively, it may be that the ratio of dark matter mass to stellar mass is higher for the Lyα–selected galaxies than for the LBGs. This possibility yields the question: where are the rest of the baryons associated with the dark matter halos of the Lyα–emitters? One simple solution is that the Lyα–emitting galaxies themselves harbor a stellar population which contributes little light to the rest–frame UV, and so has remained heretofore unobserved. Both an older ($t > 10^8$ years) population, and/or a dust–shrouded population fit this criterion. In either case, the UV contribution of the missing population would be too small to affect the rest–frame equivalent of the Lyα line in the presence of a younger, UV–bright population. As described in § 7.2, these scenarios could be disentangled with a campaign of deep, near– and mid–infrared observations.

We caution that these conclusions are based on our current photometric sample of narrowband–selected Lyα–emitting galaxy candidates. Though bolstered by the spectroscopy contained in this thesis, this sample still suffers some uncertainties. For example, the measured correlation length and volume densities of the candidates vary amongst our five narrowband filters at $z \approx 4.5$. This discrepancy could be due to varying degrees of contamination, to the different depths achieved in each filter, or to genuine large scale structure. Further spectroscopy aimed at increasing the confirmation fraction would aid us in distinguishing between these possibilities.

7.4   The History of Reionization

The spectroscopic observations of the $z > 6$ quasars yielded the first detections of the long–awaited Gunn–Peterson trough, seeming to imply that the intergalactic medium (IGM) is neutral to $z \approx 6$ (Becker et al. 2001; Djorgovski et al. 2001; Fan et al. 2002). The initial, naive interpretation of this discovery was that it heralded the direct detection of the end of the cosmic dark ages, the epoch whose death knell was sounded by the birth of the first stars and quasars with sufficient energy to reionize the IGM.

Subsequently, our picture of the reionization history of the IGM has become more complicated. As summarized in Stern et al. (2005), the *Wilkinson Microwave Anisotropy Probe* (WMAP) recently identified a large amplitude signal in the temperature–polarization maps of the cosmic microwave background (Spergel et al. 2003) indicating a large optical depth to Thomson scattering and favoring reionization at $z \approx 15$, not $z \approx 6$. The WMAP results are not necessarily inconsistent with those of the quasar Gunn–Peterson troughs, however. Only a small neutral fraction ($x_{\mathrm{HI}}^{\mathrm{IGM}} \sim 0.001$) is required to produce the Gunn–Peterson effect, so one plausible scenario is that reionization was an extended event, beginning early but not completing until $z \approx 6$.

Unfortunately, even this simple solution is contraindicated by models of the large cosmological Strömgren spheres expected to surround high–redshift quasars, which imply a larger IGM neutral fraction ($x_{\mathrm{HI}}^{\mathrm{IGM}} > 0.1$) at $z \sim 6$ than is suitable for the extended reionization scenario (Wyithe & Loeb 2003; Mesinger & Haiman 2004). Moreover, a variety of theoretical models now suggest that reionization occurred twice, first at $z \approx 20$ with the onset of zero–metallicity Population III stars, and then again by massive Population II stars after a partial recombination (e.g. Cen 2003; Haiman & Holder 2003; Somerville et al. 2003). See Barkana & Loeb (2001), Loeb & Barkana (2001), and Miralda-Escudé (2003) for recent reviews.

High–redshift Lyα–emitting galaxies offer another angle of attack on this issue, as the visibility of Lyα emission should be a sensitive function of the IGM neutral fraction. Stern et al. (2005) and Malhotra & Rhoads (2004) present first attempts to exploit this fact by attempting to track the suppression of Lyα emission across reionization as it manifests itself in Lyα–emission–line luminosity functions created at increasing slices in redshift. To date, the results of comparing Lyα luminosity functions at $z \approx 5.7$ and $z \approx 6.5$ are consistent with the null hypothesis: that there is no strong evolution in the luminosity functions, and therefore that the IGM neutral fraction does not substantially increase over the intervening 160 Myrs. However, both samples suffer from extremely small numbers; the Stern et al. (2005) result is based on just six galaxies culled from several separate surveys, each replete with its own systematics. Clearly, the sensitive implementation of this technique will require a monolithic spectroscopic effort of at least the size and scope as the campaigns we described in Chapters 5 and 6, but shifted to redshift $z \approx 6.5$ and beyond.

7.5   Results of Pilot Study

A pilot study employing the suite of coordinated observations suggested in § 7.2 has been undertaken which follows–up narrowband–selected $z \approx 5.7$ galaxy candidates in the Chandra Deep Field–South. The narrowband imaging is supplemented by both broadband optical imaging and by *Spitzer*/IRAC mid–infrared imaging provided by the Great Origins Deep Survey (GOODS; Giavalisco et al. 2004a). Four $z \approx 5.7$ candidates were identified in the optical imaging, one of which is clearly detected in the *Spitzer* 3.6 μm data. The strength of this emission (1.8 μJy), along with a vigorous star formation rate (SFR) suggested by the rest–frame UV data, implies that this object has a substantial, mature ($t > 10^8$ years) stellar mass, and is therefore not primordial (Rhoads 2005, private communication). The three remaining candidates have 3.6 μm upper limits ranging down to 0.11 μJy. Combined with the modest Lyα–derived SFRs of $\approx 7\ M_\odot$ yr^{-1}, the 3.6 μm nondetections imply that the ages of these starbursts are at most $\sim 10^7$ years (Mobasher et al., in progress). Clearly, these tantalizing initial results deserve to be bolstered, either through the selection and systematic spectroscopic follow–up of high–redshift Lyα–emitting candidates in the GOODS fields, or by deep *Spitzer* observations in fields where such catalogs already exist, such as those described in this thesis.

Bibliography

Abraham, R. G., Valdes, F., Yee, H. K. C., & van den Bergh, S. 1994, ApJ, 432, 75

Abraham, R. G., van den Bergh, S., Glazebrook, K., Ellis, R. S., Santiago, B. X., Surma, P., & Griffiths, R. E. 1996, ApJS, 107, 1

Ahn, S. 2004, ApJ, 601, L25

Ajiki, M. et al. 2002, ApJ, 576, L25

—. 2003, AJ, 126, 2091

Alexander, D. M., Brandt, W. N., Hornschemeier, A. E., Garmire, G. P., Schneider, D. P., Bauer, F. E., & Griffiths, R. E. 2001, AJ, 122, 2156

Alexander, D. M., Vignali, C., Bauer, F. E., Brandt, W. N., Hornschemeier, A. E., Garmire, G. P., & Schneider, D. P. 2002, AJ, 123, 1149

Ando, M., Ohta, K., Iwata, I., Watanabe, C., Tamura, N., Akiyama, M., & Aoki, K. 2004, ApJ, 610, 635

Antonucci, R. 1993, ARA&A, 31, 473

Aussel, H., Cesarsky, C. J., Elbaz, D., & Starck, J. L. 1999, A&A, 342, 313

Aussel, H., Gerin, M., Boulanger, F., Desert, F. X., Casoli, F., Cutri, R. M., & Signore, M. 1998, A&A, 334, L73

Baade, W. & Minkowski, R. 1954, ApJ, 119, 206

Baldwin, J. A., Phillips, M. M., & Terlevich, R. 1981, PASP, 93, 5

Barger, A. J., Cowie, L. L., Brandt, W. N., Capak, P., Garmire, G. P., Hornschemeier, A. E., Steffen, A. T., & Wehner, E. H. 2002, AJ, 124, 1839

Barger, A. J., Cowie, L. L., Capak, P., Alexander, D. M., Bauer, F. E., Fernandez, E., Brandt, W. N., Garmire, G. P., & Hornschemeier, A. E. 2003, AJ, 126, 632

Barger, A. J., Cowie, L. L., & Richards, E. A. 2000, AJ, 119, 2092

Barger, A. J., Cowie, L. L., Trentham, N., Fulton, E., Hu, E. M., Songaila, A., & Hall, D. 1999, AJ, 117, 102

Barkana, R. & Loeb, A. 2001, Phys. Rep., 349, 125

Barmby, P. et al. 2004, ApJS, 154, 97

Baron, E. & White, S. D. M. 1987, ApJ, 322, 585

Bassani, L., Dadina, M., Maiolino, R., Salvati, M., Risaliti, G., della Ceca, R., Matt, G., & Zamorani, G. 1999, ApJS, 121, 473

Baugh, C. M., Cole, S., Frenk, C. S., & Lacey, C. G. 1998, ApJ, 498, 504

Becker, R. H. et al. 2001, AJ, 122, 2850

Beers, T. C., Flynn, K., & Gebhardt, K. 1990, AJ, 100, 32

Bennett, A. S. 1962, MmRAS, 68, 163

Bertin, E. & Arnouts, S. 1996, A&AS, 117, 393

Bianchi, S., Cristiani, S., & Kim, T.-S. 2001, A&A, 376, 1

Blumenthal, G. & Miley, G. 1979, A&A, 80, 13

Bouwens, R. J. et al. 2003, ApJ, 595, 589

—. 2004, ApJ, 616, L79

Boyle, B. J. 1990, MNRAS, 243, 231

Brandt, W. N. et al. 2001, AJ, 122, 2810

Broadhurst, T. J., Ellis, R. S., Koo, D. C., & Szalay, A. S. 1990, Nature, 343, 726

Bromm, V., Kudritzki, R. P., & Loeb, A. 2001, ApJ, 552, 464

Bullock, J. S., Wechsler, R. H., & Somerville, R. S. 2002, MNRAS, 329, 246

Bunker, A. J., Marleau, F. R., & Graham, J. R. 1998, AJ, 116, 2086

Bunker, A. J., Stanway, E. R., Ellis, R. S., & McMahon, R. G. 2004, American Astronomical Society Meeting Abstracts, 204,

Bunker, A. J., Stanway, E. R., Ellis, R. S., McMahon, R. G., & McCarthy, P. J. 2003, MNRAS, 342, L47

Bunker, A. J., Warren, S. J., Clements, D. L., Williger, G. M., & Hewett, P. C. 1999, MNRAS, 309, 875

Carilli, C. L. & Yun, M. S. 1999, ApJ, 513, L13

Cen, R. 2003, ApJ, 591, 12

Chapman, S. C., Blain, A. W., Ivison, R. J., & Smail, I. R. 2003, Nature, 422, 695

Charlot, S. & Fall, S. M. 1993, ApJ, 415, 580

Cohen, J. G., Cowie, L. L., Hogg, D. W., Songaila, A., Blandford, R., Hu, E. M., & Shopbell, P. 1996, ApJ, 471, L5+

Cohen, J. G., Hogg, D. W., Blandford, R., Cowie, L. L., Hu, E., Songaila, A., Shopbell, P., & Richberg, K. 2000, ApJ, 538, 29

Cohen, J. G., Hogg, D. W., Pahre, M. A., Blandford, R., Shopbell, P. L., & Richberg, K. 1999, ApJS, 120, 171

Conselice, C. J. 1997, PASP, 109, 1251

Conselice, C. J., Bershady, M. A., & Jangren, A. 2000, ApJ, 529, 886

Cowie, L. L. & Hu, E. M. 1998, AJ, 115, 1319

Cowie, L. L., Songaila, A., Hu, E. M., & Cohen, J. G. 1996, AJ, 112, 839

Crampton, D. & et al. 2000, in ASP Conf. Ser. 207: Next Generation Space Telescope Science and Technology, 149

Cuby, J.-G., Le Fèvre, O., McCracken, H., Cuillandre, J.-C., Magnier, E., & Meneux, B. 2003, A&A, 405, L19

Danese, L., de Zotti, G., & di Tullio, G. 1980, A&A, 82, 322

Davis, M. et al. 2003, in Discoveries and Research Prospects from 6- to 10-Meter-Class
    Telescopes II. Edited by Guhathakurta, Puragra. Proceedings of the SPIE, Volume 4834,
    pp. 161-172 (2003)., 161–172

Dawson, S., McCrady, N. Stern, D., Eckart, M., Spinrad, H., Liu, M., & Graham, J. 2003,
    AJ, 125, 1236

Dawson, S., Spinrad, H., Stern, D., Dey, A., van Breugel, W., de Vries, W., & Reuland,
    M. 2002, ApJ, 570, 92

Dawson, S., Stern, D., Bunker, A. J., Spinrad, H., & Dey, A. 2001, AJ, 122, 598

Dawson, S. et al. 2004, ApJ, 617, 707

de Breuck, C. 2000, PhD thesis, University of Leiden

Dey, A., Spinrad, H., & Dickinson, M. 1995, ApJ, 440, 515

Dey, A., Spinrad, H., Stern, D., Graham, J. R., & Chaffee, F. H. 1998, ApJ, 498, L93

Dickinson, M. 1997, in AIP Conf. Proc. 470: After the Dark Ages: When Galaxies Were
    Young, ed. S. H. . E. Smith, Vol. 470 (Woodbury, New York: AIP)

Dickinson, M. 1998, in STScI Symp. Ser. 11, The Hubble Deep Field, ed. M. Livio, S. M.
    Fall, & P. Madau, Vol. 11 (New York: Cambridge University Press), 219

Dickinson, M. 2000, in Philosophical Transactions of the Royal Society of London, Series
    A, Volume 358 (London: The Royal Society), 2001

Dickinson, M. & Giavalisco, M. 2002, in ESO/USM Workshop: The Mass of Galaxies at
    Low and High Redshift (Heidelberg: Springer)

Dickinson, M. et al. 2004, ApJ, 600, L99

Djorgovski, S., Spinrad, H., McCarthy, P., & Strauss, M. A. 1985, ApJ, 299, L1

Djorgovski, S., Strauss, M. A., Spinrad, H., McCarthy, P., & Perley, R. A. 1987, AJ, 93,
    1318

Djorgovski, S. G. 1992, in ASP Conf. Ser. 24: Cosmology and Large-Scale Structure in the Universe, 73

Djorgovski, S. G., Castro, S., Stern, D., & Mahabal, A. A. 2001, ApJ, 560, L5

Dressler, A. & Gunn, J. E. 1990, in ASP Conf. Ser. 10: Evolution of the Universe of Galaxies, 200–208

Driver, S. P., Fernandez–Soto, A., Couch, W. J., Odewahn, S. C., Windhorst, R. A., Phillips, S., Lanzetta, K., & Yahil, A. 1998, ApJ, 496, L93

Eales, S. A. & Rawlings, S. 1993, ApJ, 411, 67

—. 1996, ApJ, 460, 68

Eddington, A. S. 1931, Nature, 127, 447

Egami, E. et al. 2005, ApJ, 618, L5

Eggen, O. J., Lynden-Bell, D., & Sandage, A. R. 1962, ApJ, 136, 748

Einstein, A. 1915a, Preuss. Akad. Wiss. Berlin, Sitzber, 844

—. 1915b, Preuss. Akad. Wiss. Berlin, Sitzber, 778

—. 1915c, Preuss. Akad. Wiss. Berlin, Sitzber, 799

Ellis, R., Santos, M. R., Kneib, J., & Kuijken, K. 2001, ApJ, 560, L119

Elston, R., Rieke, G. H., & Rieke, M. J. 1988, ApJ, 331, L77

Elston, R., Rieke, M. J., & Rieke, G. H. 1989, ApJ, 341, 80

Evans, A. S. 1998, ApJ, 498, 553

Faber, S. M. et al. 2003, Proc. SPIE, 4841, 1657

Fan, X., Narayanan, V. K., Strauss, M. A., White, R. L., Becker, R. H., Pentericci, L., & Rix, H. 2002, AJ, 123, 1247

Fan, X. et al. 2001, AJ, 121, 54

—. 2003, AJ, 125, 1649

Ferguson, H. C., Dickinson, M., & Williams, R. 2000, ARA&A, 38, 667

Ferland, G. J. & Osterbrock, D. E. 1986, ApJ, 300, 658

Fernández–Soto, A., Lanzetta, K. M., & Yahil, A. 1999, ApJ, 513, 34

Fomalont, E. B. 1996, in IAU Symp. 175: Extragalactic Radio Sources, ed. R. Ekers,
    C. Fanti, & L. Padrielli (New York: Kluwer Academic Publishers), 555

Fomalont, E. B., Kellermann, K. I., Richards, E. A., Windhorst, R. A., & Patridge, R. B.
    1997, ApJ, 475, L5+

Francis, P. J., Hewett, P. C., Foltz, C. B., Chaffee, F. H., Weymann, R. J., & Morris, S. L.
    1991, ApJ, 373, 465

Frei, Z., Guhathakurta, P., Gunn, J. E., & Tyson, J. A. 1996, AJ, 111, 174

Friedmann, A. 1922, Z. Phys., 10, 377

Frye, B., Broadhurst, T., & Benítez, N. 2002, ApJ, 568, 558

Fujita, S. S. et al. 2003, AJ, 125, 13

Fynbo, J. U., Möller, P., & Thomsen, B. 2001, A&A, 374, 443

Gallego, J., Zamorano, J., Rego, M., Alonso, O., & Vitores, A. G. 1996, A&AS, 120, 323

Gardner, J. P., Brown, T. M., & Ferguson, H. C. 2000, ApJ, 542, L79

Giacconi, R. et al. 2001, ApJ, 551, 624

—. 2002, ApJS, 139, 369

Giavalisco, M. et al. 2004a, ApJ, 600, L93

—. 2004b, ApJ, 600, L103

Gnedin, N. Y. & Ostriker, J. P. 1997, ApJ, 486, 581

Graham, J. R. & Dey, A. 1996, ApJ, 471, 720

Haiman, Z. & Holder, G. P. 2003, ApJ, 595, 1

Haiman, Z. & Loeb, A. 1997, ApJ, 483, 21

—. 1998, ApJ, 503, 505

Hamann, F. & Ferland, G. 1993, ApJ, 418, 11

Harrison, E. R. 1974, ApJ, 191, L51+

Hasinger, G. 1999, in MPR Rep. 272, Highlights in X–ray Astronomy, ed. B. Aschenbach
& M. Freyberg (Garching: Max–Planck Institute fur extraterretrische Physik), 199

Hasinger, G., Burg, R., Giacconi, R., Schmidt, M., Trumper, J., & Zamorani, G. 1998,
A&A, 329, 482

Heckman, T. 2000, in ASP Conf. Ser. 240, Gas & Galaxy Evolution, ed. J. Hibbard,
M. Rupen, & J. van Gorkom (San Francisco: ASP), in press, [astro–ph/0009075]

Heckman, T., Krolik, J., Meurer, G., Calzetti, D., Kinney, A., Koratkar, A., Leitherer, C.,
Robert, C., & Wilson, A. 1995, ApJ, 452, 549

Heckman, T. M., Armus, L., & Miley, G. K. 1990, ApJS, 74, 833

Heckman, T. M., Lehnert, M. D., Strickland, D. K., & Armus, L. 2000, ApJS, 129, 493

Hogg, D. W., Cohen, J. G., Blandford, R., & Pahre, M. A. 1998, ApJ, 504, 622

Hogg, D. W., Neugebauer, G., Armus, L., Matthews, K., Pahre, M. A., Soier, B. T., &
Weinberger, A. J. 1997, AJ, 113, 474

Hogg, D. W. et al. 2000, ApJS, 127, 1

Horne, K. 1986, PASP, 98, 609

Hornschemeier, A. E. et al. 2000, ApJ, 541, 49

—. 2001, ApJ, 554, 742

Hu, E. & McMahon, R. G. 1996, Nature, 382, 281

Hu, E. M., Cowie, L. L., Capak, P., McMahon, R. G., Hayashino, T., & Komiyama, Y.
2004, AJ, 127, 563

Hu, E. M., Cowie, L. L., & McMahon, R. G. 1998, ApJ, 502, L99

Hu, E. M., Cowie, L. L., McMahon, R. G., Capak, P., Iwamuro, F., Kneib, J.-P., Maihara, T., & Motohara, K. 2002, ApJ, 568, L75

Hu, E. M., McMahon, R. G., & Cowie, L. L. 1999, ApJ, 522, L9

Hu, E. M., McMahon, R. G., & Egami, E. 1996, ApJ, 459, L53+

Hu, E. M. & Ridgway, S. E. 1994, AJ, 107, 1303

Hubble, E. 1929, Proceedings of the National Academy of Science, 15, 168

Hubble, E. P. 1925, The Observatory, 48, 139

—. 1926, ApJ, 63, 236

Hughes, D. H. et al. 1998, Nature, 394, 241

Iwata, I., Ohta, K., Tamura, N., Ando, M., Wada, S., Watanabe, C., Akiyama, M., & Aoki, K. 2003, PASJ, 55, 415

Jannuzi, B. T. & Dey, A. 1999, in ASP Conf. Ser. 191, Photometric Redshifts and High Redshift Galaxies, ed. R. J. Weymann, L. J. Storrie-Lombardi, M. Sawicki, & R. J. Brunner (San Francisco: ASP), 111

Kaiser, N. 1991, ApJ, 383, 104

Kauffmann, G., White, S. D. M., & Guiderdoni, B. 1993, MNRAS, 264, 201

Kennicutt, R. C. & Kent, S. M. 1983, AJ, 88, 1094

Kleinmann, S. G., Hamilton, D., Keel, W. C., Wynn-Williams, C. G., Eales, S. A., Becklin, E. E., & Kuntz, K. D. 1988, ApJ, 328, 161

Kodaira, K. et al. 2003, PASJ, 55, L17

Koo, D. C. & Kron, R. T. 1980, PASP, 92, 537

Kuchinski, L. E., Madore, B. F., Freedman, W. L., & Trewhella, M. 2001, AJ, 122, 729

Kudritzki, R.-P. et al. 2000, ApJ, 536, 19

Kunth, D., Mas-Hesse, J. M., Terlevich, E., Terlevich, R., Lequeux, J., & Fall, S. M. 1998, A&A, 334, 11

Lacy, M. et al. 1994, MNRAS, 271, 504

Larkin, J. E., McLean, I. S., Graham, J. R., Becklin, E. E., Figer, D. F., Gilbert, A. M.,
    Levenson, N. A., Teplitz, H. I., Wilcox, M. K., & Glassman, T. M. 2000, ApJ, 533, L61

Lehmann, I. et al. 2000, A&A, 354, 35

Lehnert, M. D. & Bremer, M. 2003, ApJ, 593, 630

Leitherer, C. et al. 1999, ApJS, 123, 3

Lemaitre, G. 1927, Ann. Soc. Sci. Bruxelles, 47, 49

—. 1931, Nature, 127, 706

Levenson, N. A., Fernandes, R. C. J., Weaver, K. A., Heckman, T. M., & Storchi–
    Bergmann, T. 2001, ApJ, 557, 54

Lilly, S. J. 1988, ApJ, 333, 161

Liu, M. C., Dey, A., Graham, J. R., Bundy, K. A., Steidel, C. C., Adelberger, K., &
    Dickinson, M. E. 2000, AJ, 119, 2556

Loeb, A. & Barkana, R. 2001, ARA&A, 39, 19

Lowenthal, J. D., Koo, D. C., Guzman, R., Gallego, J., Phillips, A. C., Faber, S. M., Vogt,
    N. P., Illingworth, G. D., & Gronwall, C. 1997, ApJ, 481, 673

Madau, P. 1995, ApJ, 441, 18

Madau, P., Ferguson, H. C., Dickinson, M. E., Giavalisco, M., Steidel, C. C., & Fruchter,
    A. 1996, MNRAS, 283, 1388

Madau, P., Haardt, F., & Rees, M. J. 1999, ApJ, 514, 648

Maier, C., Meisenheimer, K., Thommes, E., Hippelein, H., Röser, H. J., Fried, J., von
    Kuhlmann, B., Phleps, S., & Wolf, C. 2003, A&A, 402, 79

Malhotra, S. & Rhoads, J. E. 2002, ApJ, 565, L71

—. 2004, ApJ, 617, L5

Malhotra, S., Wang, J. X., Rhoads, J. E., Heckman, T. M., & Norman, C. A. 2003, ApJ, 585, L25

Mann, R. G. et al. 1997, MNRAS, 289, 482

Manning, C., Stern, D., Spinrad, H., & Bunker, A. J. 2000, ApJ, 537, 65

Martin, C. L. 1998, ApJ, 506, 222

—. 1999, ApJ, 513, 156

Mas–Hesse, J. M., Kunth, D., Tenorio–Tagle, G., Leitherer, C., Terlevich, R. J., & Terlevich, E. 2003, ApJ, 598, 858

Massey, P. & Gronwall, C. 1990, ApJ, 358, 344

Massey, P., Valdes, R., & Barnes, J. 1993, A User's Guide to Reducing Slit Spectra with IRAF, available at http://iraf.noao.edu/iraf/web/docs/spectra.html

McCarthy, J. K. et al. 1998, in Proc. SPIE Vol. 3355, Optical Astronomical Instrumentation, ed. S. D'Odorico (Bellingham: SPIE), 81

McCarthy, P. J. 1993, ARA&A, 31, 639

McCarthy, P. J., Dickinson, M., Filippenko, A. V., Spinrad, H., & van Breugel, W. J. M. 1988, ApJ, 328, L29

McLean, I. S. et al. 1998, in Proc. SPIE Vol. 3354, Infrared Astronomical Instrumentation, ed. A. M. Fowler (Bellingham: SPIE), 566

Meier, D. L. & Terlevich, R. 1981, ApJ, 246, L109

Mesinger, A. & Haiman, Z. 2004, ApJ, 611, L69

Miralda-Escudé, J. 2003, Science, 300, 1904

Mo, H. J., Mao, S., & White, S. D. M. 1998, MNRAS, 295, 319

Moran, E. C., Halpern, J. P., & Helfand, D. J. 1996, ApJS, 106, 341

Moran, E. C., Kay, L. E., Davis, M., Filippenko, A. V., & Barth, A. J. 2001, ApJ, 556, L75

Mushotzky, R. F. & Scharf, C. A. 1997, ApJ, 482, L13

Nagao, T. et al. 2004, ApJ, 613, L9

Nandra, K. & Pounds, K. A. 1994, MNRAS, 268, 405

Neufeld, D. A. 1991, ApJ, 370, L85

Norman, C. et al. 2002, ApJ, 571, 218

Oke, J. B. & Gunn, J. E. 1983, ApJ, 266, 713

Oke, J. B. et al. 1995, PASP, 107, 375

Opik, E. 1922, ApJ, 55, 406

Osterbrock, D. E. 1989, The Astrophysics of Gaseous Nebulae and Active Galactic Nuclei
    (Mill Valley: University Science)

Ostriker, J. P. & Gnedin, N. Y. 1996, ApJ, 472, L63

Ouchi, M., Yamada, T., Kawai, H., & Ohta, K. 1999, ApJ, 517, L19

Ouchi, M. et al. 2003, ApJ, 582, 60

—. 2004, ApJ, 611, 660

Partridge, R. B. & Peebles, P. J. E. 1967, ApJ, 147, 868

Pascarelle, S. M., Windhorst, R. A., Driver, S. P., Ostrander, E. J., & Keel, W. C. 1996,
    ApJ, 456, L21

Pascarelle, S. M., Windhorst, R. A., & Keel, W. C. 1998, AJ, 116, 2659

Pavlovsky, C. et al. 2001, in ACS Instrument Handbook (Baltimore: STScI)

Peacock, J. A. 1999, Cosmological Physics, 1st edn. (Cambridge University Press)

Peebles, P. J. E. 1971, A&A, 11, 377

Petitjean, P. et al. 1996, Nature, 380, 411

Pettini, M., Shapley, A. E., Steidel, C. C., Cuby, J., Dickinson, M., Moorwood, A. F. M.,
    Adelberger, K. L., & Giavalisco, M. 2001, ApJ, 554, 981

Phillips, A. C., Guzman, R., Gallego, J., Koo, D. C., Lowenthal, J. D., Vogt, N. P., Faber, S. M., & Illingworth, G. D. 1997, ApJ, 489, 543

Pirzkal, N. et al. 2005, ApJ, 622, 319

Press, W. H. & Schechter, P. 1974, ApJ, 187, 425

Press, W. H., Teukolsky, S. A., Vetterling, W. T., & Flannery, B. P. 1992, Numerical Recipes in C: The Art of Scientific Computing, 2nd edn. (Cambridge University Press)

Pritchet, C. J. 1994, PASP, 106, 1052

Ravindranath, S. & Prabhu, T. P. 2001, Ap&SS, 276, 593

Rawlings, S., Lacy, M., Blundell, K. M., Eales, S. A., Bunker, A. J., & Garrington, S. T. 1996, Nature, 383, 502

Renzini, A., Ciotti, L., D'Ercole, A., & Pellegrini, S. 1993, ApJ, 419, 52

Rhoads, J. E. & Malhotra, S. 2001, ApJ, 563, L5

Rhoads, J. E., Malhotra, S., Dey, A., Stern, D., Spinrad, H., & Jannuzi, B. T. 1999, Bulletin of the American Astronomical Society, 31, 1405

—. 2000, ApJ, 545, L85

Rhoads, J. E. et al. 2003, ApJ, 125, 1006

—. 2004, ApJ, 611, 59

Richards, E. A. 2000, ApJ, 533, 611

Richards, E. A., Kellermann, K. I., Fomalont, E. B., Windhorst, R. A., & Partridge, R. B. 1998, AJ, 116, 1039

Riess, A. G. et al. 2001, ApJ, 560, 49

Rodighiero, G., Granato, G. L., Franceschini, A., Fasano, G., & Silva, L. 2000, A&A, 364, 517

Rosati, P. et al. 2002, ApJ, 566, 667

Sandage, A., Freeman, K. C., & Stokes, N. R. 1970, ApJ, 160, 831

Santos, M. R. 2004, MNRAS, 349, 1137

Scannapieco, E. & Broadhurst, T. 2001, ApJ, 549, 28

Scannapieco, E., Schneider, R., & Ferrara, A. 2003, ApJ, 589, 35

Schaerer, D. 2002, A&A, 382, 28

—. 2003, A&A, 397, 527

Schneider, D. P., Schmidt, M., & Gunn, J. E. 1994, AJ, 107, 880

Schneider, D. P. et al. 2000, AJ, 120, 2183

Schreier, E. J. et al. 2001, ApJ, 560, 127

Shapley, A. E., Steidel, C. C., Adelberger, K. L., Dickinson, M., Giavalisco, M., & Pettini, M. 2001, ApJ, 562, 95

Shapley, A. E., Steidel, C. C., Pettini, M., & Adelberger, K. L. 2003, ApJ, 588, 65

Sheinis, A. I., Miller, J. S., Bolte, M., & Sutin, B. M. 2000, in Proc. SPIE Vol. 4008, Optical and IR Telescope Instrumentation and Detectors, ed. M. Iye & A. Moorwood (Bellingham: SPIE), 522

Somerville, R. S., Bullock, J. S., & Livio, M. 2003, ApJ, 593, 616

Somerville, R. S. & Primack, J. R. 1999, MNRAS, 310, 1087

Spergel, D. N. et al. 2003, ApJS, 148, 175

Spinrad, H., Dey, A., Stern, D., & Bunker, A. 1999, in Proc. KNAW Colloq., The Most Distant Radio Galaxies, ed. H. Rottgering, N. Best, & M. Lenhert (Amsterdam: KNAW), 522

Spinrad, H. & Djorgovski, S. 1984, ApJ, 285, L49

Spinrad, H., Filippenko, A. V., Wyckoff, S., Stocke, J. T., Wagner, R. M., & Lawrie, D. G. 1985, ApJ, 299, L7

Spinrad, H., Stern, D., Bunker, A., Dey, A., Lanzetta, K., Yahil, A., Pascarelle, S., & Fernández–Soto, A. 1998, AJ, 116, 2617

Stanway, E. R., Bunker, A. J., & McMahon, R. G. 2003, MNRAS, 342, 439

Stanway, E. R., Bunker, A. J., McMahon, R. G., Ellis, R. S., Treu, T., & McCarthy, P. J. 2004a, ApJ, 607, 704

Stanway, E. R. et al. 2004b, ApJ, 604, L13

Steidel, C. C., Adelberger, K. L., Giavalisco, M., Dickinson, M., & Pettini, M. 1999, ApJ, 519, 1

Steidel, C. C., Adelberger, K. L., Shapley, A. E., Pettini, M., Dickinson, M., & Giavalisco, M. 2000, ApJ, 532, 170

Steidel, C. C., Giavalisco, M., Dickinson, M., & Adelberger, K. L. 1996a, AJ, 112, 352

Steidel, C. C., Giavalisco, M., Pettini, M., Dickinson, M., & Adelberger, K. L. 1996b, ApJ, 462, L17

—. 1996c, ApJ, 462, L17

Steidel, C. C. & Hamilton, D. 1992, AJ, 104, 941

Steidel, C. C., Pettini, M., & Adelberger, K. L. 2001, ApJ, 546, 665

Stern, D., Bunker, A., Spinrad, H., & Dey, A. 2000a, ApJ, 537, 73

Stern, D., Dey, A., Spinrad, H., Maxfield, L., Dickinson, M., Schlegel, D., & González, R. A. 1999, AJ, 117, 1122

Stern, D., Eisenhardt, P., Spinrad, H., Dawson, S., van Breugel, W., Dey, A., de Vries, W., & Stanford, S. A. 2000b, Nature, 408, 560

Stern, D. & Spinrad, H. 1999, PASP, 111, 1475

Stern, D., Tozzi, P., Stanford, S. A., Rosati, P., Holden, B., Eisenhardt, P., Elston, R., Wu, K. L., Connolly, A., Spinrad, H., Dawson, S., Dey, A., & Chaffee, F. H. 2002a, AJ, 123, 2223

Stern, D., Yost, S. A., Eckart, M. E., Harrison, F. A., Helfand, D. J., Djorgovski, S. G., Malhotra, S., & Rhoads, J. E. 2005, ApJ, 619, 12

Stern, D. et al. 2002b, ApJ, 568, 71

Stiavelli, M., Scarlata, C., Panagia, N., Treu, T., Bertin, G., & Bertola, F. 2001, ApJ, 561, L37

Stockton, A. & Ridgway, S. E. 1998, AJ, 115, 1340

Surdej, J. 1979, A&A, 73, 1

Taniguchi, Y. et al. 2003, ApJ, 585, L97

—. 2005, PASJ, 57, 165

Tenorio–Tagle, G., Silich, S. A., Kunth, D., Terlevich, E., & Terlevich, R. 1999, MNRAS, 309, 332

Teplitz, H. I., Gardner, J. P., Malumuth, E. M., & Heap, S. R. 1998, ApJ, 507, L17

Thompson, D. & Djorgovski, S. G. 1995, AJ, 110, 982

Thompson, R. I. & Storrie–Lombardi, L. J. 1997, in AIP Conf. Proc. 470: After the Dark Ages: When Galaxies Were Young, ed. S. H. . E. Smith, Vol. 470 (Woodbury, New York: AIP)

Tielens, A. G. G. M., Miley, G. K., & Willis, A. G. 1979, A&AS, 35, 153

Tody, D. 1993, in ASP Conf. Ser. 52, Astronomical Data Analysis Software and Systems II, ed. R. Hanisch, R. Brissenden, & J. Barnes (San Francisco: ASP), 173

Tumlinson, J., Shull, J. M., & Venkatesan, A. 2003, ApJ, 584, 608

Valageas, P. & Silk, J. 1999, A&A, 347, 1

van Breugel, W., De Breuck, C., Stanford, S. A., Stern, D., Röttgering, H., & Miley, G. 1999, ApJ, 518, L61

van Breukelen, C., Jarvis, M. J., & Venemans, B. P. 2005, MNRAS, 359, 895

van den Bergh, S. 2002, PASP, 114, 797

van den Bergh, S., Abraham, R. G., Whyte, L. F., Merrifield, M. R., Eskridge, P. B., Frogel, J. A., & Pogge, R. 2002, AJ, 123, 2913

Vanden Berk, D. E. et al. 2001, AJ, 122, 549

Veilleux, S. & Osterbrock, D. E. 1987, ApJS, 63, 295

Vernet, J., Fosbury, R. A. E., Villar-Martín, M., Cohen, M. H., Cimatti, A., di Serego Alighieri, S., & Goodrich, R. W. 2001, A&A, 366, 7

Waddington, I., Windhorst, R. A., Cohen, S. H., Partridge, R. B., Spinrad, H., & Stern, D. 1999, ApJ, 526, L77

Wang, J. X. et al. 2004, ApJ, 608, L21

Weymann, R. J., Stern, D., Bunker, A., Spinrad, H., Chaffee, F. H., Thompson, R. I., & Storrie–Lombardi, L. J. 1998, ApJ, 505, L95

White, S. D. M. & Rees, M. J. 1978, MNRAS, 183, 341

Williams, R. E. et al. 1996, AJ, 112, 1335

Wyithe, J. S. B. & Loeb, A. 2003, ApJ, 588, L69

Xue, Y. & Wu, X. 2000, ApJ, 538, 65

Yan, H. & Windhorst, R. A. 2004, ApJ, 612, L93

Yan, H., Windhorst, R. A., & Cohen, S. H. 2003, ApJ, 585, L93

Yan, H. et al. 2005, ApJ, in press [astro–ph/0507673]

Zepf, S. E. 1997, Nature, 390, 377

Zepf, S. E., Moustakas, L. A., & Davis, M. 1997, ApJ, 474, L1+

Zhang, Y., Anninos, P., Norman, M. L., & Meiksin, A. 1997, ApJ, 485, 496